Construction Safety Informatics

Rita Yi Man Li

Construction Safety
Informatics

 Springer

Rita Yi Man Li
Department of Economics and Finance
Hong Kong Shue Yan University
Hong Kong, China

ISBN 978-981-13-5760-2   ISBN 978-981-13-5761-9   (eBook)
https://doi.org/10.1007/978-981-13-5761-9

Library of Congress Control Number: 2019931516

This Springer imprint is published by the registered company Springer Nature Singapore Pte Ltd.
The registered company address is: 152 Beach Road, #21-01/04 Gateway East, Singapore 189721, Singapore

Acknowledgements

I would like to thank all my colleagues who co-authored with me for the chapters:

Kwong Wing Chau, Department of Real Estate and Construction, the University of Hong Kong (Chaps. 2, 4 and 6)

Daniel Chi Wing Ho, Faculty of Design and Environment, Technical and Higher Education Institute of Hong Kong (Chaps. 2, 4 and 6)

Wilson Weisheng Lu, Department of Real Estate and Construction, the University of Hong Kong (Chaps. 2, 6 and 7)

Darren Tat Ki Fung, Department of Economics and Finance, Hong Kong Shue Yan University (Chap. 2)

Frankie Fanjie Zeng, Sustainable Real Estate Research Center, Hong Kong Shue Yan University (Chap. 2)

Tat Ho Leung, Department of Global Development, University of Manchester (Chaps. 2, 5, 6 and 9)

Li Meng, University of South Australia (Chap. 5)

Mandy Wai Yee Lam, Sustainable Real Estate Research Center, Hong Kong Shue Yan University (Chap. 6)

Beiqi Tang, Sustainable Real Estate Research Center, Hong Kong Shue Yan University (Chap. 7)

and

Cho Kei Mak, Sustainable Real Estate Research Center, Hong Kong Shue Yan University (Chap. 10)

January 2019   Rita Yi Man Li

Contents

Chapter 1
Three Generations of Construction Safety Informatics: A Review

Abstract Recent developments in information technology, artificial intelligence, deep learning and related affordable technologies, such as wireless sensors, open-source software and web apps, have led to scientific breakthroughs in various fields, including the construction industry. In this chapter, we propose a new area of research, 'construction safety informatics' and introduce the idea that there have been three generations of construction safety informatics. We describe the progress of informatics, generally, in the modern era, and then its application in the construction industry. Its ontology concerning construction safety enhancement, management and data study is also reviewed. The results show that academia has mainly been concerned with the first generation of construction safety informatics which relied utterly on human control. Some researchers have started to look at the second generation, which encompasses the development of the Internet of Things (IoT) to communicate and generate the requisite data automatically. The construction industry, however, has gone one step further to develop a chatbot which can assist safety officers in filling out their safety reports.

Keywords Construction safety informatics · Information technology · Artificial intelligence

1 Introduction

Over the past two decades, rapid developments in mobile technologies, Web 2.0 channels and apps have increased information availability substantially. Technological breakthroughs, including artificial intelligence, big data analytics, open-source software such as Python and R have meant that almost everybody, even including those who are not rich in material resources, can participate in informatics practice, research and development. Likewise, the development of cost-effective hardware, such as the Raspberry Pi and sensors, provides an affordable means to develop useful tools for, for instance, construction practitioners. Also, the idea of informatics has experienced an unprecedentedly rapid development across some different areas. For example, the public has an unprecedented ability to obtain data relating to, and

© Springer Nature Singapore Pte Ltd. 2019
R. Y. M. Li, *Construction Safety Informatics*,
https://doi.org/10.1007/978-981-13-5761-9_1

to take part actively in evidence-based health care (Eysenbach and Jadad 2001). Pedoinformatics, the application of informatics approaches to soils data, has been applied in unsupervised multivariate cluster analyses recently to predict sorption and contaminant degradation in varied soils (Furey et al. 2019).

Despite a systematic cognitive sequence of safety information was established for safety science in general (Luo and Wu 2019), there is a lack of systematic review on research that falls under the umbrella of construction safety informatics. This chapter aims to propose a new field of research relating to the construction industry: construction safety informatics. The followings are topics relevant to this:

Software:

- Information, data and knowledge sharing through various information technologies;
- Data analytics on construction safety and accident records;
- Application of artificial intelligence for safety analysis, applications and image analysis;
- Big data analytics targeted at safety records; this can be useful for forecasting;
- Software developments which can improve safety performance on sites, such as Building Information Modelling software.

Hardware:

In short, that relates to items that we can see physically in the construction industry.

- Robotics can reduce the number of labourers on construction sites and the number of accidents;
- Internet of Things applications that involve sensors specifically to avoid construction accidents;
- Various technologies such as virtual reality, augmented reality, etc. which can be used for safety training;
- Various smart products such as smart safety helmets.

There are three generations of construction safety informatics:

- The first generation of construction safety informatics consisted of technologies that relied completely on control by human beings; for example, structural equation modelling requires the work of an analyst.
- The second generation of construction safety informatics included smart features such as the Internet of Things which can send information to human operators, without human intervention—from sensors, etc. Yet, these 'smart' tools cannot learn and improve on their own capabilities.
- The third generation of construction safety informatics uses state-of-the-art AI, to mimic human behaviour and think, act, learn and improve on its own decision-making. All that is required is that the relevant information is fed to these systems, so that they can be 'taught'.

Whilst the construction industry is labour intensive for many reasons, skyrocketing labour costs, coupled with huge compensation costs, drives informaticians to develop the new informatics tools to improve construction safety. Many researchers have been attempting to develop the second generation of construction safety tools such as smart helmets (which are in use on some sites). In addition, some construction companies in the US and Hong Kong are trying to develop third-generation technologies. One construction company in Hong Kong is currently using a chatbot that facilitates their safety officers to fill in safety report. This chatbot can help the safety officers to fill in the relevant items when the officers miss out information. For example, if a safety officer is asked where an accident happened by the chatbot, the safety officer may input 12 or twelve or floor 12, but the chatbot will automatically insert 12/F into the report directly.

2 Informatics: A Review on Areas Other than Construction Industry

Some have defined informatics as the science of applying information age technology to serve the specialised needs of one of several practical domains. These domains can include any category of practice (Carney and Kong 2017). The term 'informatics' attracts several differing interpretations. Informatics studies the illustration, processing and communications of information in engineered and natural systems. The central notion is the transformation of information by computation or communication, organisms or artefacts. Other definitions are linked to information and communications technology which is the science that is concerned with the gathering, manipulating, storing, retrieving and classifying of records. The discipline (of informatics) develops new uses for information technology and studies how people transform technology, and technology transforms us (Reynolds et al. 2008).

Despite there being a lack of research in academia regarding IoT, the electronic industries have already started to put forward AI-capable IoT devices—in 2018. These devices are capable of learning about their owners and allow for the implementation of a 'smart life' whereby services are optimised. Things, the cloud, AI and devices become more than just objects; they become lifestyle partners with feelings and knowledge (Sharp 2018). On 19 September 2018, https://www.limitlessiq.com/news/post/view/id/6694/.

3 Construction Informatics

Construction informatics was right off the bat proposed by Turk (2006); this publication is usually known as 'Construction IT' or 'communication and information technologies in construction'. It exhibits its metaphysics that, together with system, epistemology and axiology, constitute a new field in the construction research. In theory, the ontology characterises 'what exists'. Regarding a scientific field—what exists for a field of study, what is its territory of discourse? The given cosmology of the development informatics has the state of order.

At its highest level, it recognises (1) central topics and (2) bolster subjects. Central themes make learning related either to (a) data handling exercises, (b) correspondence/coordination exercises or about (c) necessary frameworks. These address the information exchange process and incorporate research needs, exchange, organisation and the effects of research. The metaphysics can be utilised to delineate a research territory, to plan educational programmes, to structure the motivation of a meeting, to provide catchphrases and phrases orders to bibliographic databases or learning administration when all is said in done (Turk 2006).

3.1 Ontology

Ontology refers to the method of sharing, exchanging and reusing domain knowledge (Lu et al. 2015). According to Ning et al. (2018), risk interaction value r_{1kl} represents the hazard involved with a temporary facility k due to another temporary facility lowing to the interaction flows between them. The risk interaction value r_{1kl} has a positive relationship with material flow, equipment flow and personnel flow and a negative relationship with information flow. Extremely high flows between facilities yield a high likelihood of an accident occurring along the path between them. The likelihood of an incident will be reduced if some safety information exists. Except for the interaction flows, the distance between facilities also has a positive relationship with safety risk because a longer distance will increase the number of conflicts caused by the interaction flows. If the positive and negative impacts of the interaction flow and the effect of distance on the safety risk level are considered, the assessment function related to the interaction flows, denoted by the risk interaction value r_{1kl} can be expressed as Eq. (1).

$$s_{1kl} = \frac{d_{ij}(\text{PF}_{kl} + \text{MF}_{kl} + \text{EF}_{kl})}{\text{IF}_{kl}} \tag{1}$$

When the facility k ($k = 1, 2, 3, ..., m$) is assigned to the location i ($i = 1, 2, 3, ..., n$) and the facility l ($i = 1, 2, 3, ..., m$) is assigned to the location i ($i = 1, 2, 3, ..., m$) where $m \leq n$ and $\text{IF}_{kl}, \text{PF}_{kl}, \text{MF}_{kl}, \text{EF}_{kl}$ are the material flow values for information flow, personnel flow, material flow and equipment flow between facilities k and l, respectively. An increase in movement frequency of personnel, equipment and materials raises the safety risk, and high information flow reduces the conflicts between them and lowers the accidents between the locations I and j in Euclidean distance:

$$d_{ij} = \sqrt{(x_j - x_i)^2 + (y_j - y_i)^2} \tag{2}$$

The safety risk level R1IR for all temporary facilities, considering interaction flows, is

$$S_{1IS} = \sum_{l=1}^{m} \sum_{k=1}^{m} s_{1IS} \tag{3}$$

$$S_{2IS} = s_D \cdot d_{ij} \cdot (1 - Z_{ik}) + s_{2IS}^0 \tag{4}$$

where s_D is the binary variable with a value of 1 when the facility is assigned to the location i and 0 as otherwise, and s_{2IS}^0 is the value for hazardous site waste facility k including hazardous equipment, materials and heavy equipment. The safety risk level of unoccupied location i ($i = 1, 2, ..., n$) from hazardous facility k when facility k ($k = 1, 2, ..., m$) is assigned to location j ($j = 1, 2, ..., n$) is expressed in equation:

$$S_{2IR} = \sum_{i=1}^{n} \sum_{j=1}^{n} \sum_{k=1}^{m} \frac{SHE_{ik} + SSW_{ik} + SHME_{ik}}{d_{ij}} \tag{5}$$

To summarise and compare the site risk levels, the safety risk associated with the safety concerns is then normalised in the following equations:

$$s_{1kl}^* = s_{1kl}/\max[s_{1kl}] \tag{6}$$

$$s_{2kl}^* = r_{2kl}/\max[s_{2kl}] \tag{7}$$

where max $[r_{1kl}]$ and max $[r_{2kl}]$ are the maximum values for r_{1kl} and r_{1kl}, and the safety risk for the interaction flows for the temporary facilities will be

$$S_{1kl}^* = \sum_{l=1}^{m} \sum_{k=1}^{m} s_{1kl}^* \tag{8}$$

For unoccupied locations and locations occupied by temporary facilities, the safety risk level corresponding to safety/environmental concerns is expressed as Eq. (9).

$$S_{2kl}^* = \sum_{i=1}^{n} \sum_{k=1}^{m} s_{2kl}^* \tag{9}$$

The safety risk level of a construction site layout is determined via the summation of Eqs. (8) and (9). The safety risk level of facility k is calculated to offer an example. In Fig. 4, a construction site is divided into numerous grid units.

3.2 Data Science

The advent of big data not only challenges the best computation approach but also enables innovative research across a wide range of topics and leads to the development of numerous large-data statistical approaches such as the artificial neural network (Li and Li 2018; Li 2018a). As there is an increase in the amount of data available due to the existence of mobile and real-time data provided by the various Internet of Things tools, data mining, modelling and analytics are playing an increasingly important role in construction safety. For example, we may utilise big data to predict safety performance and machine learning tools to help us analyse big data.

Shohet et al. (2018)'s simulation results show that the optimal safety investment is 1.0% of the project cost. For example, compare two parallel state-of-the-art unsupervised machine learning families of techniques, graph mining and hierarchical clustering on principal components, on an attribute data set obtained from scanning 5298 unstructured injury reports with Tixier's Natural Language Processing tool (Tixier et al. 2017).

3.3 Software and Cloud Platform

Software plays a vital role in many areas, for instance, education, record keeping and retrieval, and of course, safety management. Building Information Modelling, for example, allows construction companies to generate safety information even before construction activities commence (Li 2018c). The cloud platform and SQL databases are used by construction companies which have to deal with lots of data daily. The artificial intelligence chatbot is an application which helps officers fill in accident reports and so helps to provide safety information (Li 2018c).

3.4 Robotics and Wearable Technologies

Wearable devices can be based on some different technologies: radar, ultra-wideband (UWB), magnetic field, ultrasonic, sonar, Global Positing System (GPS), Radio-Frequency Identification (RFID), laser, Bluetooth, Electrocardiogram (ECG/EKG) and Electromyography (EMG). Sensors such as gyroscopes, Galvanic Skin Response (GSR) sensors, accelerometers and magnetometers can comprise a body sensor network (Awolusi et al. 2018). The IoT technologies, such as the metre-level RFID-based location and tracking technology, the centimetre-level ultrasonic detection technology, the infrared access technology, etc., were developed within a three-tier network architecture concerning helping workers change their risky behaviours and avoid accidents on modern construction sites (Zhou and Ding 2017). Besides, safety warning for road construction workers can also be generated with the help of IoT (Li 2018b).

The collection of data via questionnaire requires a great deal of time and effort. Additionally, some workers, especially in developing countries, are not well-educated. Accordingly, additional time is needed to clarify the survey and record their responses. Then again, interviewing is often not plausible since too many questioners are needed in order to meet with the large number of labourers who may be involved. Therefore, some researchers proposed a covert strategy for monitoring labourers' physical status by gathering their physical information, such as cardiovascular framework pointers, electrodermal markers and electrodermal indicators, via wearable devices.

3.5 Blockchain and Artificial Intelligence for Knowledge Sharing

Currently, the Internet of Things (e.g. QR code), Web 2.0 (LinkedIn, Facebook) and mobile apps can be used for construction knowledge sharing. The present author's research amongst 110 construction practitioners globally via LinkedIn (see the distribution of respondents in Table 1) in 2018 found that many people consider monetary rewards are important drivers for sharing knowledge using the abovementioned tools.

Yet, as neither the developers nor the contractors will want to spend extra money on using these tools, mass media developments based on blockchain may provide a good avenue for this kind of knowledge sharing. We expect that another wave of knowledge sharing will use Steemit. Powered by blockchain technology, Steemit uses a new cryptocurrency to reward users who upload commentary, images, articles,

Table 1 Geographical distribution of the survey respondents among the 110 responses

City	Number of respondents	City	Number of respondents
Abu Dhabi	1	Islamabad Pakistan	1
Africa	1	Jilin, Qingdao, Guangzhou	1
Bellary	1	Korea	1
Bikaner	1	Lahore	2
Cape town	1	London	1
Dubai	1	New Delhi	1
Erdos	1	Pakistan	1
Foshan	2	Shenzhen	1
Guangzhou	1	Taiwan Shanghai	1
Hangzhou	1	Taiyuan	1
Hong Kong	80	UAE	1
Hong Kong and Dubai	1	Wuhan	1

etc. Users get paid through the sourcing and up-voting of the content by other users. The earlier content is up-voted, the higher the reward. Users are paid in 'Steem Power' and in Steem Dollars, which can then be exchanged for US Dollars. Given the attractive monetary return, this tool definitely attracts some online bloggers to switch their publication to Steemit. The sharing of safety knowledge via this channel would attract the same kind of benefits (Steemit 2018).

4 Materials and Methods

The literature review includes papers that investigated the domain of safety in the construction industry. The literature search was accomplished in two steps. In the first step, titles, abstracts and keywords were searched using a manual search within databases of papers published between January 2010 and January 2018 in Science Direct. Review articles and articles that were not related to informatics were excluded as the primary aim was to locate studies related to the three generations of safety informatics.

The results show that the academics seem to be keen on the first and second generations of construction safety informatics. Many of the papers describe applied data science projects. Some of these have started to utilise the machine learning approach (an area of artificial intelligence). Yet, these models do not allow us to teach and make them able to think and make decisions (Table 2).

Note that although the research approaches remain with the second generation, machine learning is one element of these studies which is also a component of AI; this has the potential to move this kind of research to looking at the third generation of safety informatics.

4.1 The Third Generation of Safety Informatics

Some international software companies have started to explore the possibility of utilising AI to enhance construction safety. As compared to humans, AI systems work faster terms of identifying hazards and finding the best permutations of solutions that lead to the elimination of conflicts of interests on sites.

To plan for the construction of a building, 3D models need to include the architecture, structural, mechanical and electrical engineering, and the plumbing (MEP) design, and the work schedule and timing related to these concerns. One challenge is to ensure that the different models from each team do not 'clash'. In the planning and design phase machine learning, in the form of generative design, can be used to identify and mitigate clashes between the different teams who will be involved in the construction (Bharadwa 2018).

GenMEP, a generative design add-on to Autodesk Revit, focuses on the mechanical, electrical and plumbing design aspects of BIM. Once a 3D building model is

Table 2 Three generations of construction safety informatics research

Categories	Previous research	Safety informatics generation
BIM	4-dimensional (4D) construction safety planning (Choe and Leite 2017)	1st
Ontology	SQL and BIM cloud (Li et al. 2018b) Cloud-based safety information and communication system (Zou et al. 2017) BIM-based knowledge base (Li et al. 2018a)	1st
Hardware		
Safety warning	A cyber 3D tower LEGO crane (Niu et al. 2019) Wearable electroencephalography (EEG) safety helmet (Chen et al. 2016) IoT which prevent accidents (Zhou and Ding 2017)	2nd
Safety inspection	Unmanned aerial vehicle (Melo et al. 2017)	1st
Data collection	Guo et al. (2017)	1st
Data science and analytics		
Machine learning	Random forest to develop leading safety indicators (Poh et al. 2018; Tixier et al. 2017)	2nd
Simulation	Fuzzy Matter Element (FME), Monte Carlo (MC) simulation technique (Zhang et al. 2017) Monte Carlo (MC) simulation Shohet et al. (2018)	1st
Fuzzy Delphi Method (FDM) and Decision-Making Trial and Evaluation Laboratory (DEMATEL)	Bavafa et al. (2018)	1st
Bayesian network factors analysis and classification system	Xia et al. (2018b)	1st
Confirmatory factor analysis	Xia et al. (2018a)	1st
Structural equation modelling	Saunders et al. (2017), Chen et al. (2017, 2018), Wu et al. (2016)	1st
Frequency-adjusted importance index, Spearman's rank correlation, T-Test	Gunduz and Ahsan (2018)	1st
Rating scale model	Fass et al. (2017)	1st
	Ning et al. (2018)	
Infographics	Zhou et al. (2015)	1st

created in Autodesk Revit, the GenMEP add-on automatically designs the routing of the electrical system for the building model to ensure that the cables do not all end up along the same route. The software utilises machine learning to explore all the permutations of solutions, quickly generates the design alternatives, and it tests and learns from each iteration with regard to what works and what does not (Bharadwa 2018).

5 Conclusion

Informatics has developed rapidly in the healthcare sector. When we search for 'informatics' on databases such as Science Direct, Wiley, etc., many of these show prominently health informatics results. In view of the rapid development of mobile technologies, artificial intelligence, etc., we propose a new field of research in relation to construction safety, i.e. construction safety informatics. In specific terms, this consists of software for safety knowledge sharing, storage and data analysis and hardware such as robotics and wearable devices. We also review the various informatics tools that have been developed recently by throwing light on the research included in the Science Direct database.

References

Awolusi, I., E. Marks, and M. Hallowell. 2018. Wearable technology for personalized construction safety monitoring and trending: Review of applicable devices. *Automation in Construction* 85: 96–106.

Bavafa, A., A. Mahdiyar, and A.K. Marsono. 2018. Identifying and assessing the critical factors for effective implementation of safety programs in construction projects. *Safety Science* 106: 47–56.

Bharadwa, R. 2018. *AI Applications in construction and building—Current use-cases* [cited 30 July 2018]. Available from https://www.techemergence.com/ai-applications-construction-building/.

Carney, T.J., and A.Y. Kong. 2017. Leveraging health informatics to foster a smart systems response to health disparities and health equity challenges. *Journal of Biomedical Informatics* 68: 184–189.

Chen, J., X. Song, and Z. Lin. 2016. Revealing the "Invisible Gorilla" in construction: Estimating construction safety through mental workload assessment. *Automation in Construction* 63: 173–183.

Chen, Y., B. McCabe, and D. Hyatt. 2017. Impact of individual resilience and safety climate on safety performance and psychological stress of construction workers: A case study of the Ontario construction industry. *Journal of Safety Research* 61: 167–176.

Chen, Y., B. McCabe, and D. Hyatt. 2018. A resilience safety climate model predicting construction safety performance. *Safety Science* 109: 434–445.

Choe, S., and F. Leite. 2017. Construction safety planning: Site-specific temporal and spatial information integration. *Automation in Construction* 84: 335–344.

de Melo, R.R.S., D.B. Costa, J.S. Álvares, and J. Irizarry. 2017. Applicability of unmanned aerial system (UAS) for safety inspection on construction sites. *Safety Science* 98: 174–185.

Eysenbach, G., and A.R. Jadad. 2001. Evidence-based patient choice and consumer health informatics in the Internet age. *Journal of Medical Internet Research* 3 (2): e19.

Fass, S., R. Yousef, D. Liginlal, and P. Vyas. 2017. Understanding causes of fall and struck-by incidents: What differentiates construction safety in the Arabian Gulf region? *Applied Ergonomics* 58: 515–526.

Furey, J., A. Davis, and J. Seiter-Moser. 2019. Natural language indexing for pedoinformatics. *Geoderma* 334: 49–54.

Gunduz, M., and B. Ahsan. 2018. Construction safety factors assessment through frequency adjusted importance index. *International Journal of Industrial Ergonomics* 64: 155–162.

Guo, H., Y. Yu, T. Xiang, H. Li, and D. Zhang. 2017. The availability of wearable-device-based physical data for the measurement of construction workers' psychological status on site: From the perspective of safety management. *Automation in Construction* 82: 207–217.

Li, R.Y.M. 2018a. Breakthroughs in algorithms and applications for big data analysis in modern business world. *International Journal of Data Analysis Techniques and Strategies* 10 (3): 205–207.

Li, R.Y.M. 2018b. *An economic analysis on automated construction safety: Internet of things, artificial intelligence and 3D printing*. Singapore: Springer.

Li, R.Y.M. 2018c. Software engineering and reducing construction fatalities: An example of the use of Chatbot. In *An economic analysis on automated construction safety: Internet of things, artificial intelligence and 3D printing*, ed. R.Y.M. Li, 105–116. Springer Singapore: Singapore.

Li, R., and H. Li. 2018. Have housing prices gone with the smelly wind? Big data analysis on landfill in Hong Kong. *Sustainability* 10 (2): 341.

Li, M., H. Yu, H. Jin, and P. Liu. 2018a. Methodologies of safety risk control for China's metro construction based on BIM. *Safety Science*.

Li, M., H. Yu, and P. Liu. 2018b. An automated safety risk recognition mechanism for underground construction at the pre-construction stage based on BIM. *Automation in Construction* 91: 284–292.

Lu, Y., Q. Li, Z. Zhou, and Y. Deng. 2015. Ontology-based knowledge modeling for automated construction safety checking. *Safety Science* 79: 11–18.

Luo, T., and C. Wu. 2019. Safety information cognition: A new methodology of safety science in urgent need to be established. *Journal of Cleaner Production* 209: 1182–1194.

Ning, X., J. Qi, and C. Wu. 2018. A quantitative safety risk assessment model for construction site layout planning. *Safety Science* 104: 246–259.

Niu, Y., W. Lu, F. Xue, D. Liu, K. Chen, D. Fang, and C. Anumba. 2019. Towards the "third wave": An SCO-enabled occupational health and safety management system for construction. *Safety Science* 111: 213–223.

Poh, C.Q.X., C.U. Ubeynarayana, and Y.M. Goh. 2018. Safety leading indicators for construction sites: A machine learning approach. *Automation in Construction* 93: 375–386.

Reynolds, P.A., J. Harper, A.M. Jenner, and S. Dunne. 2008. Better informed: an overview of health informatics. *British Dental Journal* 204: 259.

Saunders, L.W., B.M. Kleiner, A.P. McCoy, K.P. Ellis, T. Smith-Jackson, and C. Wernz. 2017. Developing an inter-organizational safety climate instrument for the construction industry. *Safety Science* 98: 17–24.

Sharp. 2018. *AI + IoT People Oriented IoT*. [cited 20 September 2018]. Available from http://www.sharp-world.com/brand/vision/aiot/about.

Shohet, I.M., M. Luzi, and M. Tarshish. 2018. Optimal allocation of resources in construction safety: Analytical-empirical model. *Safety Science* 104: 231–238.

Steemit. 2018. *What is STEEMIT and how does it work?* [cited 13 October 2018]. Available from https://steemit.com/steemit/@panenka/what-is-steemit-and-how-does-it-work.

Tixier, A.J.P., M.R. Hallowell, B. Rajagopalan, and D. Bowman. 2017. Construction safety clash detection: Identifying safety incompatibilities among fundamental attributes using data mining. *Automation in Construction* 74: 39–54.

Turk, Ž. 2006. Construction informatics: Definition and ontology. *Advanced Engineering Informatics* 20 (2): 187–199.

Wu, C., F. Wang, P.X.W. Zou, and D. Fang. 2016. How safety leadership works among owners, contractors and subcontractors in construction projects. *International Journal of Project Management* 34 (5): 789–805.

Xia, N., M.A. Griffin, X. Wang, X. Liu, and D. Wang. 2018a. Is there agreement between worker self and supervisor assessment of worker safety performance? An examination in the construction industry. *Journal of Safety Research* 65: 29–37.

Xia, N., P.X.W. Zou, X. Liu, X. Wang, and R. Zhu. 2018b. A hybrid BN-HFACS model for predicting safety performance in construction projects. *Safety Science* 101: 332–343.

Zhang, L., L. Ding, X. Wu, and M.J. Skibniewski. 2017. An improved Dempster-Shafer approach to construction safety risk perception. *Knowledge-Based Systems* 132: 30–46.

Zhou, Z., Y.M. Goh, and Q. Li. 2015. Overview and analysis of safety management studies in the construction industry. *Safety Science* 72: 337–350.

Zhou, C., and L.Y. Ding. 2017. Safety barrier warning system for underground construction sites using Internet-of-things technologies. *Automation in Construction* 83: 372–389.

Zou, P.X.W., P. Lun, D. Cipolla, and S. Mohamed. 2017. Cloud-based safety information and communication system in infrastructure construction. *Safety Science* 98: 50–69.

Chapter 2
Modelling Construction Accident Tort Cases

Abstract This chapter sheds light on construction accident compensation via an analysis using mathematical modelling. First, we introduce the Nash equilibrium under the negligence rule. After that, we utilise the risk compensation and thermostat theory to investigate risk-taking behaviours on the sites. Furthermore, we attempt to reveal a mathematical interpretation of accident compensation. The results of this study show that (1) the negligence rule forces workers to maintain a high level of due care and minimise total social costs and (2) risk-taking behaviour is an optimal solution to workers, in that it is unlikely that taking zero risks is the best for them. As many places around the globe adopt common law in their formal institutions which govern construction compensation, the paper illustrates its practical implications for the legal and construction industries in similar jurisdictions.

Keywords Construction safety · Economic analysis · Mathematical model · Tort

1 Introduction

Construction accidents on sites lead to overwhelming social costs and losses (Li and Poon 2009, 2013). One major reason that lessens the likelihood of accidents on sites is the high expenditures of accidents. Therefore, judges' decisions on the costs of accidents play an important role in encouraging the employer to provide a safe working environment. Likewise, workers are encouraged to work safely they are liable for any accident which is due to their own reckless behaviour. Whilst previous research conducted by Li and Poon (2013) and Li (2015) studied construction accident compensation from a law and economics' perspective, Lim et al. (2018) revealed the relationship between scale and accident compensation, research on liability and negligence in tort and the risk of construction compensation undertaken from a mathematical perspective is scarce.

Given all this, an analysis of the liability and compensation relationships which exist between employers, contractors and victims has essential academic and practical implications. In economics, the relationship between various stakeholders in accident

© Springer Nature Singapore Pte Ltd. 2019
R. Y. M. Li, *Construction Safety Informatics*,
https://doi.org/10.1007/978-981-13-5761-9_2

compensation can be studied under a game theory framework. It is believed that the Nash equilibrium can be achieved under the negligence rule.

Similar to many types of occupational accident court cases, construction accident court cases fall under the umbrella of tort. According to Oxford (2017), and tort is mainly concerned with compensation for personal injury and property damage due to negligence. It also protects interests such as reputation, freedom from defamation, individual freedom, title to the property, enjoyment of property and commercial interests. For a case in tort to succeed, it must be shown that the wrong was done intentionally or negligently. Most torts are actionable if they have caused damage. The person liable is the one who committed the tort, i.e. the tortfeasor.

Nevertheless, under the principle of vicarious liability, one may be liable for the tort which is committed by another person. Examples would include a negligent taxi driver who allows a third party (another driver) to cause injury to his (the taxi driver's) passenger. The passenger may sue either the person causing the accident or the taxi driver in tort.

Finding an optimal solution for tort mainly sheds light on negligence liability, strict liability and contributory negligence. It also enables us to study the legal innovations which are feasible, such as compensation for future damage and compensation criteria (Diamond 1974a). Looking at liability studies the optimal level of care among different stakeholders in a construction project (Posner and William 1980). The economic literature on tort has pointed out that there is a close relationship between economic and tort analysis. The rational assumption is that stakeholders minimise risk and maximise their benefits. In this chapter, we analyse tort based on mathematical models. By doing so, we study the application of economic theory to tort and investigate the issue based on mathematical models to evaluate safety risks and accident compensation in the construction industry. In risk compensation theory, the risk thermostat model provides some useful insights into the issue. Individuals may have a propensity to take the risk because at a particular time they consider the possible benefits may exceed the costs; the actual mechanism in effective risk decision-making involves the so-called risk homeostasis theory (Adams 1995; Sweeting 2011; Wilde 1976).

2 Economics Analysis of the Tort

Tort provides a useful means to redistribute the negative impact of accidents and offers incentives to reduce the level of safety risks on sites. Stakeholders minimise risk and maximise gain in the compensation process, e.g. Hylton (2001) and Mau (2010). According to Dongen and Verdam (2016), the development of contributory negligence theory was slow under the strict liability rule, and the share of liability among different stakeholders was not discussed before 1945. Besides, De Mot et al. (2015) provided evidence that courts with more cases are more likely to adopt a contributory negligence approach, whereas courts with fewer cases adopt both contributory and comparative negligence approaches—when judges make decisions on construction accidents compensation. Hylton and Lin (2013) show a new model which allows for the studying of the relationship between negligence and causation. Their model suggests that the information possessed by the court plays an important role in negligence tests.

Thomas (2004) developed an economic model of liability in relation to accident tort cases and showed that only when stakeholders engage in risky activities, do they have the right to bargain with the other stakeholders regarding compensation. As such, the compensation reflects the appropriate liability assignment, irrespective to the liability rule. Gruning (2006) investigates economics loss due to various stakeholders' negligence. The study found that there might be conflicts between policies and economic loss. The study concedes that American tort has an agreement that liabilities caused by economic losses, should be set, but there is no agreement on the precise amount.

However, Bayern (2009) challenge the role of economics in tort; he believes that economic theories fail to explain negligence. Similarly, Anderson (2006) argues that economic models cannot fail to provide all the clues to different parties' liabilities outside the regime of contract, i.e. in tort. Thus, liability analysis varies case by case. Faure and Weber (2015) argue that if there is dispersed loss, this leads to financial failure due to negative externality, because of potential enforcement failure. If there is a small harm suffered by an individual as a result of a tort, s/he may not be willing to take legal action to recover it. And it is this which leads to the problem of financial failure.

3 Risk Management and Compensation in Construction

The Occupational Safety and Health Potential Risk Model (OSH-PRM) was proposed by Sousa et al. (2015). They used it to estimate the cost–benefit relationship in occupational safety and health risk management. Pinto (2014) developed a new fuzzy QRAM model which aimed to reduce safety risks at work. Instead of studying risk management on construction sites, Feng and Wu (2013) explored construction workers' compensation via case study and collected data through direct observations, questionnaires and interviews. The results showed that existing risk compensation behaviours vary because of previous experiences and whether the plaintiffs had been injured or not previously—on construction sites. The workers with more experience in compensation cases displayed lower safety awareness.

By soliciting 574 occupational fatality cases from 1999 to 2011 from the Northern Region Inspection Office of the Council of Labour Affairs of Taiwan, and then conducting Pearson correlation analyses and analyses of variance (ANOVA), Liao and Chiang (2015) found that construction accident compensation has correlations with contractor size and daily salary. In short, previous studies in the literature have mainly utilised qualitative analysis to reveal the economic issues and the liability in tort. Studies on negligence and research into mathematical models of risk compensation are relatively scarce. We aim to fill the gap between risk theory and application via mathematical modelling.

4 No Liability, Strict Liability and Negligence Rule in Tort

Apart from strict and no liability, some incidents involve liabilities which fall between these two endpoints (Jain 2015). When a no liability ruling is made, the employer does not need to pay accident compensation—because the accident is understood to have happened purely due to the employees' negligence. The notion of self-interest maximisation suggests that employers do not wish to pay for safety precautions. S/he considers a/he does not have any duty of care and hence $\varphi = 0$. Nevertheless, construction workers wish to prevent accidents (Posner and William 1980). Even when the employer cares about his employees and pays for accident precaution, $y > 0$, the level of care among the workers is higher than that of the employer $\beta^1 > \gamma^1$ (Posner and William 1980).

Under the strict liability rule, employees who cause accidents are liable for their acts. Hence, an employee has to take their own safety precautions, whereby $B_y = -p_y D$ where the level of care to a potential injurer equals zero, $\beta^* = 0$ (Posner and William 1980). Posner and William (1980) are of the view that the transaction costs in the case of strict liability exceed no liability. Under the lens of the negligence rule, no one needs to bear the compensation burden when accidents happen. Under the negligence rule, the employer has taken proper due care, but the worker has not, i.e. $\gamma = \gamma^*$ and $\beta = 0$. An accident's social cost is minimised since only the employer has to bear the cost of caring, such that no compensation is needed and the worker does not bear the cost of caring but does the positive cost of damage (Posner and William 1980):

$$L(0, \gamma^*) = p(0, \gamma^*)D + B(\gamma^*) \tag{1}$$

Here, the major concern is whether the employer can shoulder the cost of the accident compensation for his employee (Posner and William 1980). The total cost for an accident victim is the sum of the caring cost paid by the employer if the employer fails to meet his own due care, $\gamma_0 < \gamma^*$:

$$p(0, \gamma_0)D + B(\gamma_0) \tag{2}$$

The cost of taking due care without any legal compensation burden will be

$$B(\gamma^*) \tag{3}$$

An accident victim has to consider the decision based on costs and benefits and opt for an optimal solution:

$$p(0, \gamma_0)D + B(\gamma_0) \gtrless B(\gamma^*) \tag{4}$$

If the cost of taking due care exceeds the cost of not taking due care plus the accident compensation, the worker will be negligent (Posner and William 1980) such that:

$$L(0, \gamma^*) = p(0, \gamma^*)D + B(\gamma^*) < L(0, \gamma_1) = p(0, \gamma_1)D + B(\gamma_1) \qquad (5)$$

where $\gamma_1 < \gamma^*$.

The negligence rule forces the worker to take due care in order to minimise the total social cost, so as to minimise the total private cost. Posner and William (1980) ignore contributory negligence in this model.

5 Nash Analysis on Contributory Negligence

According to Lai et al. (2015), the Nash equilibrium is a useful tool for modelling the problem when there are a limited number of solutions. The Nash equilibrium helps us identify all the possible choices employers and employees, who wish to maximise their benefits and their minimise costs, can make. Posner and William (1980) suggest that equilibrium is possible to attain according to the following table.

Tables 1 and 2 show that a worker takes due care level β^* in order to minimise economic and non-economic losses due to accidents at work when s/he holds the view that her or his employer is negligent. According to Posner and William (1980), an accident's social cost is

$$L(\beta, \gamma) = p(\beta, \gamma)D + A(\beta) + B(\beta) \qquad (6)$$

Table 1 Expected loss suffered by worker in case of accidents

	Worker's level of due care	
Employer's level of due care	$\beta_0 < \beta^*$	β^*
$\gamma_0 < \gamma^*$	$p(\beta_0, \gamma_0)D + A(\beta_0)$	$A(\beta^*)$
γ^*	$p(\beta_0, \gamma^*)D + A(\beta_0)$	$p(\beta^*, \gamma^*)D + A(\beta^*)$

Table 2 Expected losses of defendant when accident occurs

	Employer's level of due care	
Worker's level of due care	$\gamma_0 < \gamma^*$	γ^*
$\beta_0 < \beta^*$	$B(\gamma_0)$	$B(\gamma^*)$
β^*	$p(\beta^*, \gamma_0)D + B(\gamma_0)$	$B(\gamma^*)$

The social cost of accident is minimised when both worker and employer take due care:

$$(\beta^*, \gamma^*) = p(\beta^*, \gamma^*)D + A(\beta^*) + B(\gamma^*) \tag{7}$$

If worker and employer take less than due care, $\gamma_0 < \gamma^*$ and $\beta_0 < \beta^*$, the worker's expected loss will be

$$p(\beta_0, \gamma_0)D + A(\beta_0) = L(\beta_0, \gamma_0) - B(\gamma_0) \tag{8}$$

The expected loss suffered by a worker if he is negligent but the contractor takes appropriate due care will be

$$p(\beta_0, \gamma^*)D + A(\beta_0) = L(\beta_0, \gamma^*) - B(\gamma^*) \tag{9}$$

Accordingly, if both parties take due care, the expected loss to the worker will be

$$p(\beta^*, \gamma^*)D + A(\beta^*) = L(\beta^*, \gamma^*) - B(\gamma^*) \tag{10}$$

If the worker takes all possible due care, but the contractor takes inadequate due care, the expected loss to the worker will be

$$A(\beta^*) = L(\beta^*, \gamma_0) - p(\beta^*, \gamma_0)D - B(\gamma_0) \tag{11}$$

Workers with a certain level of contributory negligence are optimal under the negligence rule. The same analytical deduction applies to the expected loss to the employer. The most satisfactory outcome (of any court case) to the employer would be negligence by the employee. Hence, the expected loss can be minimised as $B(\gamma_0)$.

6 Due Level of Care in Tort

Posner's risk-neutral assumption suggests that workers and contractors have a linear utility function, which is $U_0 = \beta_0, +\alpha$ where $\beta > 0$ and $\alpha \geq 0$ (Posner and William 1980). It is assumed that A is a victim of an accident and B is an employer. When an accident occurs, and the level of care provided by A is β where the standard of care taken by B is φ, the likelihood of an accident depends on the level of responsibility represented by β and φ (Posner and William 1980):

$$p = p(\beta, \gamma) \tag{12}$$

Posner and William (1980) suggest that the marginal product of care is harmful and decreases in nature, such that the level of care is negatively related to the likelihood of an accident at a diminishing rate. When we take the second-order condition, it will become

$$\frac{\partial\left(\frac{\partial p}{\partial p_\beta}\right)}{\partial p_\beta} = p_{\theta\theta} > 0 \tag{13}$$

$$\frac{\partial\left(\frac{\partial p}{\partial p_\gamma}\right)}{\partial p_\gamma} = p_{\gamma\gamma} > 0 \tag{14}$$

The model studies the expected utility of worker and employer. Assume D is the worker's monetary losses as the result of an accident (Posner and William 1980). The expected utility of A will then become

$$\overline{U^A} = p\left(I^A - D - A(\beta)\right) + (1 - p)\left(I^A - A(\beta)\right) = I^A - pD - A(\theta) \tag{15}$$

And the expected utility of B will be

$$\overline{U^B} = p\left(I^B - A(\gamma)\right) + (1 - p)\left(I^A - A(\gamma)\right) = I^B - B(\gamma) \tag{16}$$

Hence, the utility of victim and employer when an accident happens:

$$\overline{U^A} + \overline{U^B} = \left[I^A - pD - A(\beta)\right] + \left[I^B - B(\gamma)\right] = I^A + I^B - pD - A(\beta) - B(\gamma) \tag{17}$$

If the utility is maximised, social welfare is maximised. The accident's social cost, i.e. the expected loss due to accident, and costs of due care $(pD + A(\beta) + B(\gamma))$ are minimised (Posner and William 1980). An accident's social costs will become (Posner and William 1980)

$$L(\beta, \gamma) = p(\beta, \gamma)D + A(\beta) + B(\gamma) \tag{18}$$

The social costs of an accident depend on the level and the marginal cost of care, A_β and B_γ are positive and do not have the tendency to drop (Posner and William 1980). If the worker and the employer select the optimal level of due care, β^* and γ^*, the accident's social cost will be minimised. The due care level of victim and employer in the first-order condition will become

$$\frac{\partial L}{\partial \beta} = 0 \tag{19}$$

$$\frac{\partial L}{\partial \varphi} = 0 \tag{20}$$

Eventually, the due care level reaches an equilibrium where the marginal cost of care equals the reduction in expected damage to the construction worker and employee when the former must pay for the victim's accident compensation (Posner and William 1980),

$$A_\beta = -p_\beta D \tag{21}$$

$$B_\gamma = -p_\gamma D \tag{22}$$

7 Contributory Negligence Tort Model

If a worker fails to take due care ($\theta_1 < \theta^*$) to prevent accidents, he has contributory negligence. Victims at work must bear the accident's costs and share the costs of damage with their employer regardless of the strict liability rule or negligence. Thus, the employer pays less than the calculated accident compensation even when he is liable for the accident. Posner and William (1980) conjecture that there are four possible results of the liability in negligence rule when the worker has contributory negligence, where s^A and s^B represent the victim and employer's contributory negligence:

According to Table 3, the employer does not have any responsibility for damage if (1) workers and employers both take due care (x^*, y^*); (2) the worker is negligent, but the employer takes due care; or (3) both of them are negligent (case 1 to case 2). Only in the case where workers perform due care but the employer is negligent, the cost of damages shifted to the defendant (Posner and William 1980). This leads to the possibility of a multiple-party analysis which implies that an accident could involve more than two parties. Multiple-party cases may be separated according to two situations: one victim to many injuries and one injury to many victims (Jain 2015).

Table 3 Four different cases of liability in the contributory negligence model

Level of due care	Liability for construction accident
β^*, γ^*	$s^A = 1$; $s^B = 0$
$\beta_1 < \beta^*$, γ^*	$s^A = 1$; $s^B = 0$
$\beta_1 < \beta^*$, $\gamma_1 < \gamma^*$	$s^A = 1$; $s^B = 0$
β^*, $\gamma_1 < \gamma^*$	$s^A = 0$; $s^B = 1$

8   Risk Compensation Theory

An increase in safety awareness alleviates the safety risks at work. The risk thermo-stat model illustrates the way workers choose between risky and safe behaviour. It hypothesises that we all have a target risk level and measure risk according to our risk thermostat. The consequences of external environmental interventions on different individuals can be different (Adams 1995). It explains why safety interventions might produce negligible results and validated with empirical data in various settings. It states that individuals behave less cautiously when they feel safe or well-protected (Feng et al. 2017).

Individuals have different levels of perceived risk (Li 2012), and risk assessment is essential in almost all decision-making processes. Each of us has a specific target level of risk-taking propensity and everyone attempt to keep their risks at this target level. If there is disparity between perceived threat and an individual's target level of risk, the individual will spontaneously adjust the level of risk to their target level by their actions. Thus, this internal assessment is known as risk homeostasis in dynamic decision-making processes. Participants do not only modify their behaviour in response to external changes but seek to counteract these changes entirely and return to their desired risk level. Therefore, risk compensation is needed as people are not machines. Safety measures change the external conditions and environment, and risk behaviour varies with this external variation (Hedlund 2000). For example, if there is an accident which occurred the year previously, contractors will take greater due care on the construction site, leading to lower accident compensation. At the same time, the site workers assess on-site hazard when they work with higher due care after the implementation of safety enhancement policy.

Wilde first proposed the risk thermostat model and revised by Adams (Adams 1985, 1988; Wilde 1976). Nevertheless, the risk thermostat model is not an opera-tional one (Adams 1995, p. 15). On the other hand, although a great deal of research has been conducted on risk compensation and the risk thermostat theory, there is no research on a generalised mathematical interpretation based on risk compensa-tion theory. Therefore, we attempt to demonstrate the risk compensation theory as it relates to construction safety from a mathematical perspective.

Risk compensation is adopted here to study the relationships between human behaviour and legislative regulations. Examples of relevant behaviours are the behaviour of drivers after seat belt laws were implemented in Europe and the US (Adams 1994; Blomquist 1988; Crandall and Graham 1984), HIV/STD Transmission (Pinkerton 2001) and the behaviours of beer drinkers (Rogers and Greenfield 1999). Under risk compensation theory, workers on sites may have a propensity to take risks and accidents still occur. Bernoullian model can be used to evaluate the net effect of changes in risk-taking behaviours on site by estimating the overall probability of accidents (Pinkerton 2001).

Thus, the risk of an accident is

$$R = 1 - \left\{ (1 - \pi) + \pi (1 - (1 - f_0 \varepsilon))^n \right\} \tag{23}$$

where n refers to the number risky actions; π is the probability that an accident occurs; f is the proportion of risk being protected from by safety apparatus and policy and ε is the effectiveness of the safety apparatus and strategy (Pinkerton 2001).

The theory simplifies the risk-taking action as it only considers minimal mutual impacts among workers on construction sites. It assumes that there no innocent person suffers if an accident occurs. To examine the consequences under risk homeostasis conditions, it supposes an increase in construction site safety apparatus and policy. Therefore, the proportion of risk being protected will shift from f_0 to f_1. The homeostatic threshold, T_{f_0, f_1}, is the intrinsic factor which balances workers' behaviours. The theory further implies that workers would have a higher incentive to take risks after receiving construction site safety improvements. From Eq. (23) (Pinkerton 2001),

$$T_{f_0, f_1} = \frac{\ln[1 - (1 - f_0 \varepsilon)]}{\ln[1 - (1 - f_1 \varepsilon)]} \tag{24}$$

where the natural logarithm function represents the offset of risk reduction from construction site safety improvement from f_0 to f_1. Thus, actual risk-taking activities increase from n to $n T_{f_0, f_1}$. Plural rationalities can influence the individual's propensity for risk-taking. The theory implies a reasonable situation whereby it is necessary for workers to cooperate to accomplish a task. The collective perception of risk on construction sites is mutually affected. From Eq. (23) (Pinkerton 2001),

$$R = 1 - \left\{ (1 - \pi) + \pi (1 - (1 - f_0 \varepsilon))^n \right\}^m \tag{25}$$

$$R = 1 - \left\{ (1 - \pi) + \pi (1 - (1 - f_1 \varepsilon))^n \right\}^m \tag{26}$$

where m shows the level of mutual risk-taking.

Site safety improvement enhances collective awareness of risk. Assume that a workers' work is fixed, and workers may increase or decrease their level of cooperation due to the nature of the task. Although the outcome varies, we can generalise the level of cooperation under the risk compensation situation τ as follows:

$$\tau = m \frac{\ln(1 - \pi) + \pi [1 - (1 - f_0 \varepsilon)]^n}{\ln(1 - \pi) + \pi [1 - (1 - f_1 \varepsilon)]^n} \tag{27}$$

The risk compensation evaluation provides a practical, scientific and significant estimation for contractors. Before safety improvement programmes are implemented, contractors can use this model to estimate the risk compensation effect among workers. It helps us to derive optimal resource allocation on-site safety (Pinkerton 2001).

9 Conclusions

Construction accident compensation is one of the best means of deterring negligent behaviour by workers, and it motivates employers to provide a safe working environment. One of the critical roles of the judge in tort cases is to determine the level of liability as between worker and employer. Many of the previous studies on construction accidents torts cases are restricted to the economic and/or legal aspects. This paper sheds light on construction accident compensation based on a mathematical and economic perspective. It fills the academic gap displayed by risk theory regarding the mathematical model analysis of construction accident compensation.

The mathematical model shows that the negligence rule suggests that the employee must take due care to minimise the total social cost which results from an accident. Having said that, however, if the price of taking due care is more than the cost of not taking sufficient due consideration, the worker will be negligent. The Nash analysis of the contributory negligence equilibrium illustrates that a worker with a certain level of contributory negligence attains the optimal solution. Finally, the risk thermostat model shows the putative mechanisms of the decision made between risk and safety behaviours. It suggests that the worker has their target risk level and measures risk, according to their risk thermostat. Thus, even though different scholars propose these theories, they share common ground—that a certain level of contributory negligence is the optimal solution. Besides, it also illustrates that judges in courts with more cases tend to adopt a contributory negligence approach, while courts with fewer cases take a contributory and/or comparative negligence rule when they make decisions regarding construction accident compensation.

Acknowledgements This chapter is a modified version of the Li, Rita Yi Man, Chau, Kwong Wing; Poon, Sun Wah; Ho, Daniel Chi Wing; Lu, Weisheng Wilson; Fung, Darren; Zeng, Frankie; Leung, Tat Ho (2017) Construction accident tort cases: mathematical and economic modelling approach 53rd Associated Schools of Construction, hosted by University of Washington and Washington State University, Seattle. We thank for the right to republish the revised version.

References

Adams, J. 1985. *Risk and freedom: The record of road safety regulation*. London: Transport Publishing Projects.

Adams, J. 1988. Evaluating the effectiveness of road safety measures. *Traffic Engineering and Control* 344–352.

Adams, J. 1994. Seat belt legislation: The evidence revisited. *Safety Science* 18: 135–152.

Adams, J. 1995. *Risk*. London: UCL Press.

Anderson, J.M. 2006. The missing theory of variable selection in the economic analysis of tort law. *SSRN Electronic Journal*. https://doi.org/10.2139/ssrn.987957.

Bayern, S. 2009. The limits of formal economics in tort law: The puzzle of negligence by Shawn Bayern. *Brooklyn Law Review* 75: 707.

Blomquist, G. 1988. *The Regulation of motor vehicle and traffic safety*. Boston: Kluwer.

Crandall, R.W., and J.D. Graham. 1984. Automobile safety regulation and offsetting behavior: Some new empirical estimates. *American Economics Review* 74: 328.

De Mot, J., Faure, M., and Klick, J. 2015. Appellate caseload and the switch to comparative negligence. *International Review of Law and Economics* 42: 147–156. https://doi.org/10.1016/j.irle.2015.01.003.

Diamond, S. 1974a. Single activity accidents. *The Journal of Legal Studies* 3 (1): 107–164.

van Dongen, E.G.D., and Verdam, H.P. 2016. The development of the concept of contributory negligence in English common law. *Utrecht Law Review* 12(1): 61. https://doi.org/10.18352/ulr.326.

Faure, M., and Weber, F. 2015. Dispersed losses in tort law – an economic analysis. *Journal of European Tort Law* 6 (2): 163–196. https://doi.org/10.1515/jetl-2015-0012.

Feng, Y., and Wu, P. 2013. Risk compensation behaviours in construction workers' activities. *International Journal of Injury Control and Safety Promotion* 22 (1): 40–47. https://doi.org/10.1080/17457300.2013.844714.

Feng, Y., P. Wu, G. Ye, and D. Zhao. 2017. Risk-compensation behaviors on construction sites: Demographic and psychological determinants. *Journal of Management in Engineering* 33 (4): 04017008.

Gruning, D. 2006. Pure economic loss in American tort law: An unstable consensus. *The American Journal of Comparative Law* 54: 187–208. https://doi.org/10.2307/20454536.

Hedlund, J. 2000. Risky business: Safety regulations, risk compensation, and individual behavior. *Injury Prevention* 6: 82–89.

Hylton, K.N. 2001. The theory of tort doctrine and the restatement of torts. Boston University School of Law Research Paper 07: 1–21.

Hylton, K.N., and Lin, H. 2013. Negligence, causation, and incentives for care. *International Review of Law and Economics* 35: 80–89. https://doi.org/10.1016/j.irle.2013.04.004.

Jain, S.K. 2015. *Economic analysis of liability rules*. India: Springer.

Lai, J.-F., Hou, J., and Wen, Z.C. 2015. A smoothing method for a class of generalized Nash equilibrium problems. *Journal of Inequalities and Applications* 90: 1–16. https://doi.org/10.1186/s13660-015-0607-6.

Li, R.Y.M. 2012. Econometric modelling of risk adverse behaviours of entrepreneurs in the provision of house fittings in China. *Construction Economics and Building* 12 (1): 11.

Li, R.Y.M. 2015. *Construction safety and waste management: An economic analysis*. Switzerland Springer.

Li, R.Y.M., and S.W. Poon. 2009. Workers' compensation for non-fatal accidents: review of Hong Kong court cases. *Asian Social Science* 5 (11): 15–24.

Li, R.Y.M., and S.W. Poon. 2013. *Construction safety*. Berlin: Springer.

Liao, C.W., and Chiang, T.L. 2015. The examination of workers' compensation for occupational fatalities in the construction industry. *Safety Science*, 72: 363–370. https://doi.org/10.1016/j.ssci.2014.10.009.

Lim, S., A.-R. Oh, J.-H. Won, and J.-J. Chon. 2018. Improvement of inspection system for reduction of small-scale construction site accident in Korea. *Industrial Health*, Article ID 2018-0033.

Mau, S.D. 2010. Tort of Negligence. In *Tort law in Hong Kong: An introductory guide*. Hong Kong University Press.

Pinkerton, S.D. 2001. Sexual risk compensation and HIV/STD transmission: Empirical evidence and theoretical consideration. *Risk Analysis* 21 (4): 727–736.

Pinto, A. 2014. QRAM a qualitative occupational safety risk assessment model for the construction industry that incorporate uncertainties by the use of fuzzy sets. *Safety Science* 63: 57–76. https://doi.org/10.1016/j.ssci.2013.10.019.

Posner, R.A., and M.L. William. 1980. The positive economic theory of tort law. *Georgia Law Review* 15: 851–924.

Rogers, J.D., and T.K. Greenfield. 1999. Beer drinking accounts for most of the hazardous alcohol consumption reported in the united states. *Journal of Studies on Alcohol* 60: 732–739.

Sousa, V., Almeida, N.M., and Dias, L.A. 2015. Risk-based management of occupational safety and health in the construction industry – part 2: Quantitative model. *Safety Science* 74: 184–194. https://doi.org/10.1016/j.ssci.2015.01.003.

Sweeting, P. 2011. *Risk assessment. In financial enterprise risk management*. Cambridge: Cambridge University Press.

Thomas, J.M. 2004. *An Economic Model of Tort Law. In The economic approach to law*. Stanford University Press.

Wilde, G.J. 1976. The risk compensation theory of accident causation and its practical consequences for accident prevention. In Annual Meeting of the Osterreichische Gesellschaft Fur Unfallchirurgies, Salzburg.

Chapter 3
Mechanisms of Safety Risk Consciousness as Reflected in Brain and Eye Activities: A Conceptual Study

Abstract Colour perception problems can impair the ability to recognise various construction safety risks on sites, while the awareness of safety signage may be affected by semiotics. This chapter first provides a review of the causes of construction accidents; this is followed by a study on the implication of colour on safety risk awareness and the impact of semiotics for safety signage. It proposes the application of mouse tracking, eye tracking, EEG and Functional Near-Infrared Spectroscopy (fNIRS) for studying hazard identifications made by workers.

Keywords Eye tracking · Semiotics · Construction safety · Integrated information theory · Construction hazard visualisation · Electroencephalogram · Functional near-infrared spectroscopy · Augmented reality · Mixed reality

1 Construction Hazard Visualisation

Albeit the construction industry is highly risky, many workers have poor hazard identification awareness (Alarcon et al. 2016). Lack of safety education and knowledge has been considered an essential cause of accidents in the industry (Li 2015a, b). Studying the construction process visually can improve the safety consciousness of workers so that they more readily understand safety management-related knowledge and the potential safety problems involved (Guo et al. 2017). Their responses can also provide useful information on safety course design.

The modified domino theory suggests that organisational factors are essential in shaping safety systems (Adam 1976). Alarcon et al. (2016) summarise the essential elements relating to construction safety management in studies published in the literature from 1976 to 2016. These identify twelve possible factors, three of which are widely agreed to be the most significant in survey results; these are safety training, management commitment and worker involvement.

There are different models and tools which have been constructed to assess construction safety awareness. For instance, Garrett and Teizer (2009) built their Human Factors Analysis Classification System (HFACS) based on a framework focused on Human Error Awareness Training (HEAT) to evaluate the safety awareness of work-

© Springer Nature Singapore Pte Ltd. 2019
R. Y. M. Li, *Construction Safety Informatics*,
https://doi.org/10.1007/978-981-13-5761-9_3

ers involved in construction projects. Externally, the management safety commitment is another well-recognised factor in construction safety awareness enhancement (Flin et al. 2000). Additionally, particularly concerning young construction workers, construction safety awareness can be enhanced in more by a teaching paradigm than by a learning paradigm (Laberge et al. 2014).

Safety training has been recognised as the main precondition for sufficient safety awareness; there are several studies in the literature which support the significance of the correlation between safety training and working behaviours (Burke et al. 2006, 2011). For example, Salas and Cannon-Bowers (2001) support that the content of safety training has been found to be crucial in increasing safety engagement among trainees.

There are different approaches in safety training, such as action regulation theory (Hacker 2003) and behavioural learning theories (Geller 2001); this indicates that such training programmes do not provide merely information but also frame the behavioural pattern of the trainees (Brahm and Singer 2013).

Although both behaviour and knowledge can be enhanced by safety training, however, in evidential terms, the outcome of such training is somewhat uncertain after eliminating the self-selection bias (Brahm and Singer 2013). Thus, the impact and effectiveness of safety training are still controversial (Pouliakas and Theodossiou 2013).

Besides, there are multiple possible factors which can potentially influence construction safety awareness; there are several studies which provide evidence regarding composite elements research. In Australia, regarding the funding for safety training issue, the industry associations, as agents, suffer from an agency related problem, that funds are misallocated into unintended areas such membership expansion (Bahn and Barratt-Pugh 2012a).

Regarding organisational structures, the agency problem is such that that the industry has become reluctant, in an unanticipated way, to engage in construction safety awareness while associations generally perceive that the training and construction safety-related support that they provide are sufficient (Bahn and Barratt-Pugh 2012b). However, the new Construction Induction Training (CIT) code, which is mandated by the Australian government, has had an impact on awareness which is positive (Bahn and Barratt-Pugh 2012b).

In Chile, according to Alarcon et al. (2016), practices which involve rewards do incentivise safety awareness while observation and monitoring have slightly negative impacts on it. In New Zealand, a series of structural equation models have been developed which emphasise the importance of production pressures being exerted towards safety awareness (Guo et al. 2016). In China, similarly, safety commitment and employee involvement are also noted as important factors for enhancing construction safety culture (He et al. 2016). In Hong Kong, according to Tononi (2015), an evaluation of the Proactive Construction Management System (PCMS) has taken place which was based on the Behavioural-Based Safety (BBS) approach which can provide significant improvements in safety performance and knowledge.

2 Impact of Colour on Safety-Related Vision and Hazard Awareness

The cognitive risks associated with construction workers relate to their recognition of hazard signals around the perilous environments. Warning signs on sites are essential tools which provide safety information, influence construction workers' behaviour and serve as reminders (Chen et al. 2018). Hence, the colour of safety signs should be chosen so that they raise workers' hazard awareness.

Among all the forms of information which can be transmitted, colour one such which significantly affects people's awareness. Ng and Chan (2015) conducted a study that examined the impacts of sign referents. Various safety sign referents were drawn up by a group of Chinese construction workers in Hong Kong. They were requested to describe their drawings and ideas. With a total of 276 illustrations, 79% were drawn with one colour, 17% in two colours, 2% in three colours and the unusual third colour was always used for letters, words and numerals. The most colour that the construction workers used most frequently was black (96%), which was followed by red (16%), blue (5%), yellow (0.36%), purple (0.36%), grey (0.36%) and green (0.36%). It was expressed that the reason for using different colours was to draw people's attention to the content of the signs. For instance, red was related to fire, a slash, a skull and an exclamation mark, while yellow was used to represent the diagonal lines of a barrier.

Furthermore, the colours in hazard signs make them easier for construction workers to easy to understand, regarding the level of hazard. Red leads to the averting of motivation. This is because of a basic instinct that humans have, which makes them feel that red signs represent something more dangerous than signs of other colours, such as white, do. Utilising an arrow as a visual direction signal allows people to effortlessly identify the direction being indicated without having to make any specify judgment. Therefore, the visualisation of a hazard or information relating to a risk using an arrow as well as colour is an effective way to enhance the recognition of a dangerous situation (Kim et al. 2017).

Different colours in safety signage transmit various messages to workers. According to the experiments in Chen et al. (2018), the colour of a safety sign is associated with the extent to which workers will visually identify the sign as relating to a hazard. For example, safety signs which include the colour red will be more easily as compared with signs which include other colours, e.g. yellow or black colour, but not red. The experiment found that red-coloured signs do not have a significant effect concerning reducing accidents, but they are easier to recognise by workers as safety signs, visually (Chen et al. 2018).

Furthermore, it has been shown that larger exit sign sizes and greater colour contrasts in public area resulted in stronger visual influences on pedestrians and can increase the probability that signage is spotted and recognised (Zhang et al. 2017). Additionally, signage colour contrast which denotes the difference between the colour of the signage and that of the surrounding environment, the difference

Table 1 Symbolic meanings related to colours

Colour	The symbolic meaning of the colour
Red	Blood, fire, a slash, a cross, a skull, an exclamation mark and a rounded shape (triangular or circular) Prohibition/fire equipment (Ng and Chan 2015) The highest level of risk (danger) (Chen et al. 2018)
Blue	Water, slippery surface, mandatory action (Ng and Chan 2015)
Purple	Chemicals (Ng and Chan 2015)
Green	A safe condition, a human figure and circular bounded shape (Ng and Chan 2015)
Grey	Chemicals (Ng and Chan 2015)
Yellow	Warning, diagonal lines of a barrier (Ng and Chan 2015)
Black	Caution (Chen et al. 2018)

between the background colour and content colour of the signage, was found to be significant factor that affects the people's awareness of the signage (Table 1).

Olander et al. (2017) researched evacuation in the case of a fire in a building. The research was based on the result of a questionnaire study which examined the design of emergency signage. 46 participants were mainly students, and 23 years old on average. It was pointed out that the utilisation of the findings should take into account the fact colours can have different meanings in different cultures. For instance, red, rather than green is adopted for emergency exit signs in the United States. Meanwhile, it was deemed that significant colour contrasts should be used when designing emergency signage as this enables the emergency signage to have good sensory affordance. It was anticipated that visual affordance would increase notably when dynamic features were adopted, like flashing lights. At the same time, visual affordance is presumed to rely strongly on the signage's ability to be conspicuous in the environment. Moreover, the signs with no red flashing lights were discarded as they were thought to have very low sensory affordance (Olander et al. 2017).

3 Semiotics and Safety Signage

Some unsafe behaviours and some accidents in the construction industry are related to insufficient hazard awareness among workers. Hence, safety signage is used to warn the workers about hazardous working environments. For example, slogans and posters are posted on sites as safety warnings (Li 2018).

Different designs of safety signs provide different levels of warning to the workers. Chen et al. (2018) developed an objective measurement paradigm by using three objective indicators, Counting Accuracy (CA), Response Time (RT) and Identification Accuracy (IA), to evaluate the cognitive effectiveness level of various safety sign designs included in an experiment. To compare the effectiveness of different shapes for the experiment, three shape groups (round, triangle and square) were selected. The experimental result showed that the forms (oval, triangle and square) of the safety signs do not affect the level of safety warning achieved by a sign. However, the round-shaped groups performed slightly better regarding CA, RT and IA which meant their round-shaped signs result in higher counting accuracies and a faster response times and workers found them more accessible to identify. Besides, the square-shaped groups performed slightly less well in IA which means square-shaped signs are more difficult to locate. Concerning CA and RT, triangle-shaped signs and square-shaped sign perform similarly (Chen et al. 2018).

Apart from the shapes of the safety signs, the content of the sign also affects the level of warning that they affect. To compare the effectiveness of different sign contents in an experiment involving three types of sign content (symbol, graphics and text) was conducted by Chen et al. (2018). The experiment found that there is a faster Response Time (RT) about the signs with graphic designs, which means that this group of signs is easier for the workers to notice. However, regarding CA and IA, graphic groups perform similarly to the other groups. Also, the signs with text and symbolic designs achieved similar results (similar result in CA, RT and IA) across the overall experiment (Chen et al. 2018).

4 Mechanisms of Consciousness: Integrated Information Theory

Integrated Information Theory (IIT) provides a mathematical explanation for the quantity and quality of conscious experience which can be applied to all known states of consciousness (Tononi 2004, 2008; Tononi and Koch 2008; Gallimore 2015; Tononi et al. 2016). In contrast, Cerullo (2015) argues that IIT fails to measure consciousness as the information exclusion it proposes is unjustified and integrated information is not sufficient for consciousness. This section attempts to demonstrate the latest mathematical formulation of IIT expressing how awareness varies over time. With a mathematical formulation, it can be easier to assess and quantify the state of consciousness.

Integrated information (φ) refers to 'the amount of information generated by a complex of elements, above and beyond the information generated by its parts' (Tononi 2008, p. 216). Specifically, 'The integrated information theory (IIT) of consciousness claims that, at the fundamental level, consciousness is integrated information' (Tononi 2008, p. 217). According to the latest version of IIT (IIT 3.0) suggested by Oizumi et al. (2014), there are several axioms and postulations which can be stated as the pillars of the model; an axiom refers to self-evident truth, and a postulate is an unproven assumption. IIT (i) information is specific among experience; (ii) integration is unified which is fundamental to non-interdependent components; and (iii) exclusion has its unique borders and a specific spatiotemporal grain. With the postulations, the Maximally basic Conceptual Structure (MICS) is consolidated; moreover, the elements which generate MICS constitute a complex. Hence, MICS specifies the quality of experience and integrates the quantity of information, ϕ^{max}.

Specifically, with regard to Oizumi et al. (2014), in terms of the mechanism of IIT, the mechanism generates meaningful information only if it constrains the states of the system's causes and effects which induces its cause–effect repertoire, while the more selective it is, the higher the level of the cause–effect information, cei, which is specified by the mechanism. After that, it is necessary to assess the irreducibility of cei. According to a simple interpretation, the arrangement has to determine whether the information is integrated. When there is one foremost cause–effect repertoire, φ^{max}, which implies the maximum value of integration or irreducibility, it is, thereby, its Maximally Irreducible Cause–Effect repertoire (MICE). Hence, the mechanism constitutes a concept once Maximally its Basic Cause–Effect repertoire (MICE) is fulfilled.

Potentially, a system of two binary elements forms the set of states with equal probability, p. At time $t = 0$, the potential repertoire, which is represented by the maximum entropy, is formally expressed as

$$p(X_0(\max H))$$

With specified states as output, we can exclude the inferior state of the input element set which constructs the actual repertoire through the mechanism and state output. Hence,

$$p(X_0(\text{mech}, x_i))$$

Therefore, in general, with effective information, i.e. generated within a system characterised by certain mechanisms in a particular state, relative entropy, H, can be measured between potential and actual repertoires:

$$ei(X(\text{mech}, \ x_i)) = H[p(X_0(\text{mech}, \ x_i))\|p(X_0(\max H))]$$

Henceforth, the next step is to assess the integration of information, or, basically, how much the data is generated by a single entity (Tononi 2008). This process can be measured by assessing the difference between the repertoires generated within the system, $p(X_0(\text{mech}, x_i))$. The actual collection of the system x with the probability distribution created by the parts considered independently:

$$\Pi_p(^k M_0(\text{mech}, \ \mu_i))$$

While $^k M$ refers to the actual repertoire of the specific component, i. Hence, the integrated information, φ, is expressed as

$$\varphi(X(\text{mech}, \ x_i)) = H\left[p(X_0 9\text{mech}, x_i)\middle\|\Pi_p(^k M_0(\text{mech}, \ \mu_i))\right] > 0 \text{ for } ^k M_0 \in \text{MIP}$$

while MIP refers to the minimum information partition which decomposes the system into minimal parts, when the information is generated in the causal interaction among its components. Hence, the system can generate integrated information, φ.

Based on the above formulation, this can be further projected to n elements. According to (Oizumi et al. 2014; Tononi 2015), with the precondition of the determined Transition Probability Matrix (TPM) of the candidate set, it specifies the probability under the external constraints. A physical system can be represented in a discrete time random vector of size n, $n \geq 2$ as

$$X_t = \{X_{1,t}, \ldots, X_{n,t}\}$$

with the observed state at time t. This implies all possible states of X_t in metric space, Ω_x, d, as

$$x_t = (x_{1,t}, \ldots, x_{n,t}) \in \Omega_x$$

Hence, the mechanism is formed within a physical system consisting of a Maximally Irreducible Cause–Effect repertoire (MICE) over a subset of elements. Let $P(X_t)$ be the power set of all possible subsets of a physical system X_t such that the candidate mechanism and past-and-future purviews are represented as

$$Y_t \in P(X_t) \text{ and } Z_{t\pm1} \in P(X_{t\pm1})$$

Thus, the cause-and-effect repertoires can be further represented, respectively, as

$$p_{\text{cause}}(z_{t-1}|y_t) \equiv \frac{1}{K} \prod_{i=1}^{|y_t|} p_{\text{cause}}(z_{t-1}|y_{i,t}), \quad z_{t-1} \in \Omega_{Z_{t-1}}$$

$$p_{\text{effect}}(z_{t+1}|y_t) \equiv \prod_{i=1}^{|z_{t+1}|} p_{\text{effect}}(z_{t+1,i}|y_t)$$

where $K = \sum_{z \in \Omega_{Z_{t-1}}} \prod_{i=1}^{y_t} p_{\text{cause}}(z|y_{i,t})$ is the normalisation constant to ensure the sum of the repertoires equals one. Hence, the cause–effect repertoire over purviews can be structured as

$$p_{\text{ce}}(y_t, Z_{t\pm1}) = \{p_{\text{cause}}(z_{t-1}|y_t), p_{\text{effect}}(z_{t+1}|y_t)\}$$

With all the cause–effect repertoires as informed by all mechanisms of a physical system in a state, the cause–effect structure can be formalised as

$$S_{\text{ce}}(x_t) = \{p_{\text{ce}}(y_t, Z_{t\pm1})|\forall(y_t, Z_{t\pm1}) \in M(x_t)\}$$

where $M(x_t)$ refers the set of all mechanisms of system X_t in a state x_t.

Meanwhile, to assess the degree of causality of a cause–effect repertoire, it is necessary to measure the cause–effect information, cei. The cause-and-effect information of candidate mechanism Y_t in a state y_t are defined as

$$ci(y_t, Z_{t-1}) = \mathrm{emd}(p_{\mathrm{cause}}(z_{t-1}|y_t), p_{\mathrm{cause}}(z_{t-1}|\emptyset), d)$$

$$ei(y_t, Z_{t+1}) = \mathrm{emd}(p_{\mathrm{effect}}(z_{t+1}|y_t), p_{\mathrm{cause}}(z_{t+1}|\emptyset), d)$$

Hence,

$$cei(y_t, Z_{t\pm1}) = \min(ci(y_t, Z_{t-1}), ei(y_t, Z_{t+1}))$$

If $cei(y_t, Z_{t\pm1}) > 0$, it implies the cause–effect power in the subset of element in a state $Y_t = y_t$.

The integrated information, φ, measures the irreducibility of a cause–effect repertoire. With the partition set P of Y_t and Z which assumes to be independent, φ is expressed as

$$\varphi(y_t, Z_{t\pm1}) = \min\left(\varphi_{\mathrm{cause}}^{\mathrm{MIP}}, \varphi_{\mathrm{effect}}^{\mathrm{MIP}}\right)$$

where

$$\varphi_{\mathrm{cause}}^{\mathrm{MIP}}(y_t, Z_{t-1}) = \min_{P} \varphi_{\mathrm{cause}}(y_t, Z_{t-1}, P)$$

and

$$\varphi_{\mathrm{effect}}^{\mathrm{MIP}}(y_t, Z_{t+1}) = \min_{P} \varphi_{\mathrm{effect}}(y_t, Z_{t+1}, P).$$

Hence, the Maximally Irreducible Cause–Effect repertoire (MICE) is $\varphi = \varphi^{\mathrm{max}}$, $\mathrm{MICE}(y_t) = p_{\mathrm{ce}}(y_t, Z_{t\pm1})$, such that $Z_{t\pm1} \neq Z_{t\pm1}^* \in P(X_{t\pm1})$, $\varphi(y_t, Z_{t\pm1}^*) \leq \varphi^{\mathrm{max}}(y_t, Z_{t\pm1})$.

5 Possible Means to Track Hazard and Safety Signage Awareness

5.1 Electroencephalogram and Eye Tracking

Specifically, several technological breakthroughs are applicable to the construction industry for the enhancing of safety awareness. For example, the Electroencephalogram (EEG), mouse tracking and eye tracking methods are innovative tools for accurately evaluating the safety practice relating to a construction project. However, these methods have never been applied in practice in the construction industry.

Despite this, it is still of value to discuss their potential benefits. Eye tracking focuses on the target physical object with which a worker is about to engage. Thereby, a useful human–machine interface, the results from the composite application of EEG and the eye tracking method, would be capable of distinguishing between a human's intentional and their spontaneous eye movements (Shishkin et al. 2016). The on-site safety management can make use of EEG tracking to monitor potential misconduct when a worker is active. Generally, applications of this composite method have been found to be able to track a user's behavioural pattern (Kim et al. 2015; Shishkin et al. 2016). Thus, such a system can systematically establish regular operation routines which the worker can be reminded of remotely if they omit any safety-related activities. A method involving combined eye tracking and mouse tracking has the additional benefit of having a similar performance with lower cost (Lupu et al. 2013).

5.2 Functional Near-Infrared Spectroscopy

EEG measures are subject to numerous artefacts due to head and body movements, and PET and fMRI require the subjects to lie down and be immobile during data acquisition. Thus, there is a need for a more robust form of measurement. Functional Near-Infrared Spectroscopy (fNIRS) is a new brain imaging technique that meets such measurement requirements with the added advantages of being field-deployable and portable. It measured the deoxygenated (HHb) and oxygenated (HbO2) haemoglobin in the brain's blood supply, and this varies with different levels of mental effort. Moreover, fNIRS allows in vivo imaging in natural conditions with natural freedom of movement within complex environments such as exist inside flight simulators (Causse et al. 2017). fNIRS neuroimaging can be used to study the activities of the working memory, short- and long-term memory, as well as various brain functions when a person sees a hazard or other things of interests (Jahani et al. 2017).

5.3 Mouse Tracking

Furthermore, in addition to the methods mentioned above, there is another innovative method based on mouse tracking called the Prospective Memory method (PM) (Abney et al. 2015). In Hehman's study, several techniques were developed to examine the onset and timing of an evolving decision process, or to test the competition between response alternatives at different time points, and to assess movement complexity alongside a spatial disorder analyses (Hehman et al. 2014). Practically, this method can be used to focus on in the evaluation of delayed behaviour within an (otherwise) regular operational routine. Therefore, it can manage the careless practice in operational activities.

5.4 Augmented Reality and Mixed Reality

Augmented Reality (AR) systems have been introduced to illustrate the progress/status of construction site activities using colour codes. For example, green indicates that activity is on-schedule, dark green means an event is ahead of schedule and red designates a process that is behind schedule (Omar et al. 2018). This colour coding system suffers from the problem that it is not suitable for use by those staff suffering from certain kind of colour blindness.

Another method which is worth to study is Mixed Reality (MR). McJunkin et al. (2018) stated that a Mixed Reality (MR) headset enables the three-dimensional visualisation of interactive holograms anchored to specific points in physical space. A mixed reality platform was developed via Unity C# programming to produce interactive three-dimensional holograms that could be displayed in the HoloLens headset. While their research was mainly intended to provide a combination of semiautomatic and manual segmentation of three-dimensional soft tissue, plus images of the bony anatomy of cadaver heads including temporal bones from two-dimensional computed tomography images, a similar approach could be used for tracking workers' awareness.

As human errors may happen about hazard identification, artificial visual intelligence which identifies the relevant objects could help. We could utilise the synthetic intelligence object detection approach, whereby safety risk photographs could be used to train a computer system about safety concerns [not the cup, etc. as illustrated by Microsoft (2018)] and the mixed reality system developed could be used for reminding workers about hazard on sites, raising workers' hazard awareness before they start working, etc.

6 Conclusion

The ability to recognise safety risks on sites can be impaired by colour, and the awareness of safety signage may be affected by semiotics. Studying the visual construction process can improve the safety consciousness of workers. Warning signs on sites are essential tools which provide safety information, influence construction workers' behaviour and serve as reminders. The use of colour in signage makes it easier for construction workers to understand the risk level being indicated, while the shapes of safety signs do not affect the level of risk that they convey.

Regarding safety risk awareness tracking, we could use mouse and eye tracking methods, or EEG or augmented reality. If we utilise the artificial intelligence object detection approach, whereby safety risk photos are used to train a computer system about safety concerns, the mixed reality presentation systems which have been developed can be used for reminding the workers about hazards on sites, and to raise workers' hazard awareness before they start working. The effects of user factors and sign referent characteristics should be considered in participatory construction safety sign redesign.

Acknowledgements This chapter is an extended and revised version of the paper published Li, Rita Yi Man, Tat Ho Leung and Tommy Au (2018) Biometrics analysis on construction workers' hazard awareness, IOP Conf. Series: Materials Science and Engineering 365, pp. 1–7.

References

Abney, D.H., D.M. McBride, A.M. Conte, and D.W. Vinson. 2015. Response dynamics in prospective memory. *Psychonomic Bulletin & Review* 22 (4): 1020–1028.

Adam, E.E. 1976. Accident causation and the management system. *Professional Safety* 21 (10): 26–29.

Alarcon, L.F., D. Acuna, S. Diethelm, and E. Pellicer. 2016. Strategies for improving safety performance in construction firms. *Accident Analysis and Prevention* 94: 107–118.

Bahn, S., and L. Barratt-Pugh. 2012a. Emerging issues of health and safety training delivery in Australia: Quality and Transferability. *Procedia—Social and Behavioral Sciences* 62: 213–222.

Bahn, S., and L. Barratt-Pugh. 2012b. Evaluation of the mandatory construction induction training program in Western Australia: Unanticipated consequences. *Evaluation and Program Planning* 35: 337–343.

Brahm, F., and M. Singer. 2013. Is more engaging safety training always better in reducing accidents? Evidence of self-selection from Chilean panel data. *Journal of Safety Research* 47: 85–92.

Burke, M.J., S.A. Sarpy, K. Smith-Crowe, S. Chan-Serafin, R.O. Salvador, and G. Islam. 2006. Relative effectiveness of worker safety and health training methods. *American Journal of Public Health* 96 (2): 315–324.

Burke, M.J., R.O. Salvador, K. Smith-Crowe, S. Chan-Serafin, A. Smith, and S. Sonesh. 2011. The dread factor: How hazards and safety training influence learning and performance. *Journal of Applied Psychology* 96 (1): 46–70.

Causse, M., Z. Chua, V. Peysakhovich, N. Del Campo, and N. Matton. 2017. Mental workload and neural efficiency quantified in the prefrontal cortex using fNIRS. *Scientific Reports* 7 (1): 5222.

Cerullo, M.A. 2015. The problem with Phi: A critique of integrated information theory. *PLoS Computational Biology* 11 (9): 1–12.

Chen, J., R.Q. Wang, J. Lin, and X. Guo. 2018. Measuring the cognitive loads of construction safety sign designs during selective and sustained attention. *Safety Science* 105: 9–21.

Flin, R., K. Mearns, P. O'Connor, and R. Bryden. 2000. Measuring safety climate identifying the standard features. *Safety Science* 34: 177–192.

Gallimore, A.R. 2015. Restructuring consciousness: The psychedelic state in light of integrated information theory. *Frontier in Human Neuroscience* 9: 1–16.

Garrett, J.W., and J. Teizer. 2009. Human factors analysis classification system relating to human error awareness taxonomy in construction safety. *Journal of Construction Engineering and Management* 135 (8): 753–763.

Geller, S.E. 2001. Behavioral-based safety in industry: Realizing the large-scale potential of psychology to promote human welfare. *Applied and Preventive Psychology* 10 (2): 87–105.

Guo, B.H.W., T.W. Yiu, and V.A. Gonzalez. 2016. Predicting safety behavior in the construction industry: Development and test of an integrative model. *Safety Science* 84: 1–11.

Guo, H., Y. Yu, and M. Skitmore. 2017. Visualization technology-based construction safety management: A review. *Automation in Construction* 73: 135–144.

Hacker, W. 2003. Action regulation theory: A practical tool for the design of modern work processes? *European Journal of Work and Organizational Psychology* 12 (2): 105–130.

He, Q., S. Dong, T. Rose, H. Li, Q. Yin, and D. Cao. 2016. Systematic impact of institutional pressures on safety climate in the construction industry. *Accident Analysis and Prevention* 93: 230–239.

Hehman, E., Stolier, R. M., and Freeman, J. B. 2014. Advanced mouse-tracking analytic techniques for enhancing psychological science. *Group Processes & Intergroup Relations* 18 (3): 384–401.

Jahani, S., A.L. Fantana, D. Harper, J.M. Ellison, D.A. Boas, B.P. Forester, and M.A. Yücel. 2017. fNIRS can robustly measure brain activity during memory encoding and retrieval in healthy subjects. *Scientific Reports* 7 (1): 9533.

Kim, M., B.H. Kim, and S. Jo. 2015. Quantitative Evaluation of a low-cost noninvasive hybrid interface based on EEG and eye movement. *IEEE Transactions on Neural Systems and Rehabilitation Engineering* 23 (2): 159–168.

Kim, K., H. Kim, and H. Kim. 2017. Image-based construction hazard avoidance system using augmented reality in wearable device. *Automation in Construction* 83: 390–403.

Laberge, M., E. MacEachen, and B. Calvet. 2014. Why are occupational health and safety training approaches not effective? Understanding young worker learning processes using an ergonomic lens. *Safety Science* 68: 250–257.

Li, R.Y.M. 2015a. *Construction safety and waste management: An economic analysis*. Switzerland Springer.

Li, R.Y.M. 2015b. Construction safety knowledge sharing via smart phone apps and technologies. In *Handbook of mobile teaching and learning*, ed. Y. Zhang, 1–11. Berlin, Heidelberg: Springer Berlin Heidelberg.

Li, R.Y.M. 2018. *An economic analysis on automated construction safety: Internet of things, artificial intelligence and 3D printing*. Singapore: Springer.

Lupu, R.G., F. Ungureanu, and V. Siriteanu. 2013. Eye tracking mouse for human computer interaction. Paper read at E-Health and Bioengineering Conference, at Iasi, Romania.

McJunkin, J.L., P. Jiramongkolchai, W. Chung, M. Southworth, N. Durakovic, C.A. Buchman, and J.R. Silva. 2018. Development of a mixed reality platform for lateral skull base anatomy. *Otology & Neurotology* 39 (10): e1137–e1142.

Microsoft. 2018. *Mr and Azure 310: Object detection* [cited 14 October 2018]. Available from https://docs.microsoft.com/en-us/windows/mixed-reality/mr-azure-310.

Ng, A.W.Y., and A.H.S. Chan. 2015. Effects of user factors and sign referent characteristics in participatory construction safety sign redesign. *Safety Science* 74: 44–54.

Oizumi, M., L. Albantakis, and G. Tononi. 2014. From the phenomenology to the mechanisms of consciousness: Integrated information theory 3.0. *PLoS Computational Biology* 10 (5): e1003588.

Olander, J., E. Ronchi, R. Lovreglio, and D. Nilsson. 2017. Dissuasive exit signage for building fire evacuation. *Applied Ergonomics* 59: 84–93.

Omar, H., L. Mahdjoubi, and G. Kheder. 2018. Towards an automated photogrammetry-based approach for monitoring and controlling construction site activities. *Computers in Industry* 98: 172–182.

Pouliakas, K., and L. Theodossiou. 2013. The economics of health and safety at work: An interdiciplinary review of the theory and policy. *Journal of Economic Surveys* 27 (1): 167–208.

Salas, E., and J.A. Cannon-Bowers. 2001. The science of training: A decade of progress. *Annual Review of Psychology* 521 (1): 471–499.

Shishkin, S.L., Y.O. Nuzhdin, E.P. Svirin, A.G. Trofimov, A.A. Fedorova, B.L. Kozyrskiy, and B.M. Velichkovsky. 2016. EEG negativity in fixations used for gaze-based control: Towards converting intentions into actions with an eye-brain-computer interface. *Frontier in Neuroscience* 10: 1–20.

Tononi, G. 2004. An information integration theory of consciousness. *BMC Neuroscience* 5 (42): 1–22.

Tononi, G. 2008. Consciousness as integrated information: A provisional manifesto. *Biological Bulletin* 215 (3): 216–242.

Tononi, G. 2015. Integrated information theory. *Scholarpedia* 10 (1): 4164.

Tononi, G., and C. Koch. 2008. The neural correlates of consciousness. *Annals of the New York Academy of Sciences* 1124 (1): 239–261.

Tononi, G., M. Boly, M. Massimini, and C. Koch. 2016. Integrated information theory: From consciousness to its physical substrate. *Nature Reviews Neuroscience* 17: 450–461.

Zhang, Z., L. Jia, and Y. Qin. 2017. Optimal number and location planning of evacuation signage in public space. *Safety Science* 91: 132–147.

Chapter 4
Dynamic Panel Study of Building Accidents

Abstract The construction industry is one of the riskiest sectors of production; it suffers from numerous deaths and injuries every year. As large amounts of compensation are spent on construction accidents, it is necessary to review the factors which impact the amount of compensation in court. In academic terms, analysis of accident compensation has usually depended on studies provided by the labour department in the past, none of the previous researches in this area has utilised the dynamic panel technique to study the relationships involved. The court case reports recorded that most victims' monthly salary fell below HK$20,000 and they aged between 30 and 49. The results show that non-Cantonese speakers receive less compensation than locals do and more extensive hearings are associated with a more significant amount of compensation. There was, however, no significant difference between victims who had and did not have a mental disorder.

Keywords Construction accidents · Court cases · Discrimination · Hong Kong · System-GMM

1 Introduction

The construction industry is one of the primary indicators of the economic performance of a country. Economic booms are usually related to high levels of construction output. Although its economic benefits are clear and it offers a multitude of job opportunities, the construction industry, in general, has a poor safety record around the globe (Irumba 2014). In some developing nations, such as Saudi Arabia, more than 50% of the total injuries at work are related to the construction industry (Panuwatwanich et al. 2017). In Europe, even though only 10% of the population is employed in the construction industry, 30% of fatal industrial accidents are related to the construction industry. In the US, the incidence rate of accidents in the construction industry is twice the industrial average. The US National Safety Council (NSC) reported that there are nearly 2200 deaths and 220,000 disabling injuries each year. In Japan, construction fatalities account for 30–40% of fatal industrial accidents and

© Springer Nature Singapore Pte Ltd. 2019
R. Y. M. Li, *Construction Safety Informatics*,
https://doi.org/10.1007/978-981-13-5761-9_4

50% in Ireland. In the UK, injuries reported in the construction sector numbered 3677 in 2005/6 as compared to 4386 in 1999/2000 and 3768 in 2004/5 (Irumba 2014).

In Hong Kong, one primary cause of construction fatalities is falling from a height (Wong et al. 2016). The number of industrial casualties in the construction industry was 19, higher than the 16 which occurred in 2006. The fatality rate per 1000 workers in the construction industry in 2015 was 0.2, which was smaller than the rate of 0.3 in 2006 and less than the average of the past 5 years by 30%. The construction industry has 39 per 1000 workers accident rate. This is higher than the overall accident rate across all sectors, which stands at around 18 per 1000 workers. Despite there being a drop in the accident rate per 1000 workers by 25% from 64 in 2006 to 39 in 2015, the number of accidents rose from 3400 in 2006 to 3723 in 2015. Moreover, the construction industry still recorded the highest accident and fatality rates among all sectors (Occupational Safety and Health Branch Labour Department 2016).

In response to the poor safety records of construction sites, various safety measures have been put into place. For example, Site Safety Working Cycle (SWC) is used for improving site safety performance. The implementation of the SWC has also been found useful in enhancing the safety awareness of frontline workers: identifying potential hazards and facilitating safety-related communications. The Independent Safety Audit Scheme (ISAS) is applied in capital works contracts involving unconventional construction methods and mega capital works contracts (contract sums exceeding HK$1000 million) (Secretary for Sustainable Development 2015). Li and Poon (2013) record that the government's attempt to improve safety on sites by introducing safety promotion and occupational health and safety assessment series 18001 has, indeed, decreased accident rates.

The labour department has jointly organised the construction industry safety award along with significant stakeholders in the industry to strengthen occupational safety and health awareness in the construction sector and elevate safety culture on sites since 1999 (Labour Department 2016). Li and Poon (2009a) indicate that recognising safety concepts and behaviours motivates workers to work safely. Safety goal acceptance and the relationships between workers are less important factors regarding safety motivation.

2 Factors Which Lead to Construction Accidents

Working on construction sites is a risky occupation. Some research studies have been carried out to identify the associated risks and to contrive mitigation strategies. Among construction trades, roofers are found to be at the highest high risk of fatality and injury, as they work at high elevations and are subject to the risk of falling (Mistikoglu et al. 2014). Also, a difference in accident rates is noted between developing and developed countries, with higher accident rates usually found in the former due to poorer safety awareness and safety measures (Li 2015).

Failures usually cause accidents by people, technology, or a combination of both. The causes are seldom singular or straightforward but are complex constellations of events, existing preconditions and system properties. There are numerous causes of accidents in the construction industry. In general, the origins of accidents can be summarised into eight categories: lack of proper training in recognising and avoiding job hazards, lack of safety equipment, deficient enforcement of safety standards, unsafe site conditions, dangerous methods of work and/or poor planning of project activities, poor attitude of workers towards safety, workers not using the provided safety equipment, and isolated sudden deviation of a worker from prescribed behaviour (Irumba 2014).

Saracino et al. (2015) found that human factors, workers' behaviours and risk perceptions, are major factors that affect the likelihood of accidents. The Hong Kong Commissioner for Labor identified worker behaviour as the major cause of construction accidents. Behaviour-based safety is an effective approach to managing safety issues (Li et al. 2015a, b). It is argued that safety motivation is an important role which affects safety behaviour. As safety motivation refers to an individual's willingness to exert effort to enact safety behaviour, employees should be made to be motivated to participate in safety activities and to work safely. An important tool for improving workers' motivation is a reward distribution system. In line with Li and Poon (2009a) and Saracino et al. (2015), it lowers the occupational accidents' occurrence rate.

By using a questionnaire survey, Feng and Wu (2015) evidenced that construction workers behave less cautiously when additional risk control measures are implemented. The effect of protective measures may be counteracted by workers' resultant risky behaviour when they perceive that they better protective measures have been put in place. Also, construction accident compensation is affected by workers' demographics: workers with higher education, more experience, or who have never been injured at work before having a greater tendency to receive accident compensation than others. Although it is noted that more experienced and educated workers tend to have better safety performance, they are more likely to receive accident compensation relating to their activities. Hence, contractors need to assess the potential influence of workers' accident compensation behaviour when they evaluate risk control measures on construction sites. Supervisors have to pay more attention to workers who have never been injured and have a higher educational level when they implement new safety measures on construction sites.

To reduce the damages payable for occupational accidents, most industrialised countries have put legislation in place to prevent work-related diseases and occupational accidents. Employers' and workers' incentives improve workplace safety and hence affect the chance of occupational-related diseases and injuries (Liao and Chiang 2015); thus incentives can lead to lower accident compensation costs accordingly.

3   Costs of Construction Accidents

The International Labour Organization (ILO) estimated that the cost of work-related illnesses and injuries amounts to 4% of the gross national product of a country. Many of the previous studies show that work-related diseases and occupational accidents are under-compensated. Previous research used workers' compensation to examine the risk of injury construction workers were subject to. By using the information available about workers' compensation, safety risk related to construction work can be assessed. Some studies have found that occupation, industry, legal counsel, union membership and healthcare costs are associated with claim costs (Liao and Chiang 2015).

In Australia, more than 2000 people die due to their work every year, and this incurs significant economic, social and personal costs. The latter reflects the fact that more than 5000 close friends and family members of workers become the survivors of traumatic work-related deaths (every year). A workers' death means that their family members children, spouse or other dependents lose income and financial support, and experience grief and other non-monetary forms of suffering (Li and Poon 2009b, 2013; Li 2015a, b; Quinlan et al. 2015). Indeed, the consequences can be dire as traumatic work-related death mostly occurs in industries such as fishing, farming, forestry, construction and road transport, where average earnings are usually low, and therefore family budgets are often tight.

In Australia and many other countries, most of the workers who die are bread-winners, and hence their family members need to rely on the workers' compensation scheme. Some families can also claim on life insurance. Furthermore, while 71% of Australian workers aged 15 or above were covered by superannuation in 2007, the median account was quite low: AUS\$31,252 for males and AUS\$18,489 for females. Workers' compensation schemes support medical and funeral expenses, and expenses for non-dependent family members who wish to attend funerals. A certain level of income is paid to the dependent family members as financial support in the case of death. The workers' compensation policy provides financial support to injured workers and their families as a means of encouraging them to return to work. The scheme aims to provide fair compensation which reduces social and economic costs. The plan requires employers to engage in no-fault insurance cover which is designed for employees who are hired under a contract of service (Quinlan et al. 2015).

In Australia and New Zealand, many self-employed workers are excluded from the workers' compensation scheme although the self-employed constitute from 15 to 17% of the active workforce. It is also important to note that there are also cases where workers themselves are reluctant to make claims. Workers are currently facing the serious challenge represented by the growth of precarious employment and employment which is temporary or undocumented. The flexibility of work arrangements and multiple job holding are often linked to high incidences of fatalities, and this makes resolving claims even harder. In many cases, there is widespread under-reporting and failure to lodge or to eventually succeed in workers' compensation claims despite supporting evidence, and even though these cases involve death or serious injuries. For example, from 2008 to 2009, of 400 families of workers who sustained a fatal injury at work, only 276 received workers' compensation while the families the other 124 did not (Quinlan et al. 2015).

Another option is available regarding securing monetary redress following death or injury at work, in Australia and some other countries. One is primarily for the families of those excluded from workers' compensation, which means most self-employed workers, in the case of traumatic work-related death; it is possible to pursue a claim for damages. The claim should follow common law relating to breach of contract or be based on a tort of negligence against the organisation or person held responsible for the worker's death. Different from the workers' compensation scheme, this is a fault-based remedy where there is no specific level of entitlement, but this is decided case by case based on the general rules (Quinlan et al. 2015).

Apart from damages claims or workers' compensation at common law, families may receive financial support through a social security system or via voluntary donations by the employer, the workmates of the deceased, the union, or the community. Previous research also shows that currently there is a significant degree of cost shifting from workers' compensation to social security systems when workers are seriously injured. It is unknown whether a similar pattern applies to the families of fatally injured workers, and not much research on the financial impacts on families suggests it does, thereby reflecting that there is a heavy burden on social security by those families denied workers' compensation. There are some notable exceptions to the general circumstances described above, as a result of, for example, funds established following workplace disasters, and employer, workmate, union and community-based funds, which are surely valuable but not a significant source of financial support (Quinlan et al. 2015).

Non-employee work arrangements and the concentrations of self-employed workers in some dangerous industries, such as forestry, fishing, farming, construction and road transport are growing; however, workers' compensation is still the most significant source of financial support for the families of deceased workers in Australia. Although the common law damages option will be studied, especially because it is important in the industries being considered here, as a result of the situations described above, workers' compensation is still the main focus of this article. It is worth noting that the employers pay part of the total costs of work-related illness and death through compensation premiums. For example, between 2008 and 2009, Australian employers paid about 16% of the total costs associated with work-related disease, injury and fatalities (compensation); this 16% amounted to AUS$6.5 billion in workers' compensation premiums. The community had to bear 10% of the total costs and workers and their families were required to carry the remaining 74%, close to three-quarters of the total costs.

Concerning the cases of death or disability due to construction accidents at work, the imbalance in the burden regarding the actual average costs to workers, employers, the community and the families falls even more heavily on the latter group (Quinlan et al. 2015). The results from a pilot study found that surviving families commented that the correlation between workers' experiences and their corresponding work compensation was negative. Those traumatically bereaved by workplace death have to face an additional burden due to the systemic issues (Quinlan et al. 2015).

4   Is Construction Accident Compensation Another Example of a Black Swan? Our Three Conjectures

Taleb et al. (2009) suggested that black swan events are almost impossible to predict and do not support the illusion that we can anticipate the future. Delay is common as many courts have large backlogs of cases (Rock and Kitty 2015). Nevertheless, the use of delay is a common strategic activity to increase the chance of success among the wealthier party as the weaker party may give up when they begin to run out of financial support. For example, in the case of lawsuits between employer and union, long delays make it more difficult for the union to retain the loyalty of the workers as such delays increase the employers' chances of dissuading employees from supporting the issue, and increases the chances of significant turnover in the workforce. A study conducted between 1952 and 1972 estimated a 0.29% drop in the chances of a labour union's victory for every one additional day added to the median average processing time (Roomkin and Block 1981).

H1: Longer processing times favour the employer more as the latter group is usually wealthier. The result of delay may be that the employers end up paying less in accident compensation.

A lack of direct physical damage lasting after an injury, and the resultant suffering being 'only' in the form of unexplained manifestations or prolonged nervous troubles does not in itself warrant the complete discontinuance of all compensation benefits, according to the leading case of Rialto Lead and Zinc Company et al. v. Industrial Commission et al. (Garve 1935). Nowadays, pain and suffering appear to be a legalistic tautology, and opinions speak of mental distress and mental illness as something different from pain and suffering or suffering. Psychological suffering may include fright, fear, terror, apprehension and anxiety. And suffering refers to something besides the mental distress which accompanies the shadow of physical pain. McCormick suggested that physical pain and mental suffering should be bracketed together like the elements of damages in personal injury cases and that physical pain is usually accompanied by psychological distress, and that the difficulty of distinguishing between these two problems has been a valid reason for allowing extra damages for mental anguish (Proehl 1961).

In Illinois, physical pain is not only a sensation which might be accompanied by psychological suffering; it is also legally recognised as a source of mental distress—from personal injuries. Hence, the judge held that '*mental suffering was compensable...we cannot readily understand how there can be the pain without mental suffering. It is a mental emotion arising from a physical injury. It is the mind that either feels or takes cognisance of physical pain, and hence there is mental anguish or suffering inseparable from bodily injury... The mental anguish which would not be proper to be considered is where it is not connected with the bodily injury, but was caused by some mental conception not arising from the physical injury'.* Nevertheless, in reality, it should be noted that a psychological phenomenon can be induced without physical injury or any sensation of physical pain. For example, close friends/relatives of workers who fall from a height may have mental illness (Proehl 1961). Hence, our second hypothesis is as follows:

H2: In practice, there can be no discrimination between mental health problems and physical suffering at the time a judge decides on the amount of compensation concerning an accident in Hong Kong.

Discrimination against foreigners is pervasive around the globe. Economists have shown their interest in discrimination since Becker's research in 1957. In the 1970s and 1980s, there was an active debate about whether prejudice should be described by Becker's 'disaster-based discrimination model, or by Arrow and Phelps' statistically based discrimination model (Guryan and Charles 2013). Discrimination has been advanced as a vivid explanation for persistent stumbling blocks encountered by ethnic minorities (Alanya et al. 2015). Zhao and Biernat (2017) used a field experiment and a lab experiment to study how White Americans reacted to foreigners who use their Anglo names or their original names. Based on self-categorisation theory, the results showed that the use of Chinese names led to fewer responses when the foreign subjects requested graduate training or appointments than using Anglo names (Zhao and Biernat 2017).

H3: Non-Cantonese speakers may have a disadvantage when seeking compensation.

5 Research Method

By using the keyword 'construction accidents' in HKILL, we attempted to discover the factors which affect accident compensation. After identifying the relevant court cases, we then used the content analysis method to extract and evaluate the occurrences of textual material with similar content in a systematic way (Li and Li 2013). Content analysis has been used in other research areas such as looking at sustainable building finance in Latin America (Li and Tsoi 2014), the line of reasoning in competition law (Li et al. 2015b) and smart home analysis (Li 2013; Li et al. 2016). We analysed the court cases according to the nature of the work, the worker's age, and the number of days between the final decision and the date of the accident.

6 Data and Research Method

We searched through the court cases available in the open-access Hong Kong Legal Information Institute (HKLII) e-database, 2016. The court cases which were decided between 2005 and 2015 took, on average, 4628 days to settle each (counted from the day of the accident). The mean amount of compensation was HK$2,631,580. The average monthly salary is HK$13,052.2 (Tables 1 and 2), and since there are missing values for the monthly salary, we used Markov Chain Monte Carlo simulation (MCMC) to deduce these.

Multiple imputations are often used to fill in missing data to generate complete sets of data. The MCMC approach consists of imputations and then later steps. Simulated values are produced to replace missing observation values, independently. The two steps iterate until a stationary distribution is reached (Yin et al. 2016).

The static panel data have endogeneity, heteroscedasticity and serial correlation problems, but fortunately, the GMM-system panel analysis provides a method to solve these problems. Arellano and Bond resolved these econometric problems in 1991; Arellano and Bover in 1995 and Blundell and Bond in 2000 developed the sys-

Table 1 Victims' age and income

Age groups	Number of workers	Income before accidents	Number of workers
17–19	2	Less than 10,000	20
20–29	9	10,000–19,999	45
30–39	22	20,000 or more	9
40–49	22		
50–59	14		
60–69	4		

Table 2 Summary statistics of the court cases

Variable	Mean	Std. dev.	C.V.	Skewness	Ex. kurtosis
Total compensation	2,631,580	16,647,700	6.32614	9.42817	87.2631
Number of days for a court case to settle (start from the date of accidents)	4628.33	11,789.7	2.54728	1.97285	6.52189
Mental suffering	0.100000 (total five workers)	0.303046	3.03046	2.66667	5.11111
Non-Cantonese speaker	0.0612245 (total six workers)	0.240974	3.93591	3.66040	11.3986
Monthly earnings before the accident (MCMC)	12,431.5	4351.69	0.350054	0.478722	−0.130948
Before accidents monthly incomes	13,052.2	4644.91	0.355870	0.237421	−0.439687

tem: the GMM estimator (GMM-SYS) along with the first difference GMM (GMM-DIF) estimator for dynamic panel data models. The GMM-SYS estimator is a system that contains both the levels and the primary difference equations. The GMM-SYS estimator provides an alternative to the standard first difference GMM estimator (Leitão and Shahbaz 2013).

The first-order autoregressive panel model is suggested by Bun and Windmeijer (2009).

7   Results and Discussion

Most of the court cases studied do not mention the educational background of the victims. About 13 had studied in secondary schools. Nevertheless, most of the plaintiffs whose educational experience is known had not finished secondary school. Two workers had completed only primary school education. Many of the court cases do not reveal the age of the victims. Among those who showed their age, many of fell into the age group 30–49.

Table 3 shows the relationships between the total amount of compensation paid on a claim and various factors based on our conjectures. A longer time is associated with a higher amount of compensation. Although it is reasonable to consider that this can provide more time for contractors to find a better position, the results go against our conjecture. This may be because a more prolonged time also implies a more complicated or severe case, so that the court needs extra time to gather suitable evidence. Court cases which have a longer time span seem to have a positive relationship with the amount of compensation paid. Besides, there is no significant relationship between workers who have a mental disorder and the total amount of compensation paid. If a plaintiff receives an accident compensation payment 1 year, this has a negative and significant impact on their subsequent year's salary. Finally, non-Cantonese speakers receive less compensation (as indicated by the substantial negative figures).

Table 3 Summary of the GMM-SYS model

Dependent variable: total compensation			
	Coefficient	Std. error	z
l_Total (−1)	−0.117245***	0.0424397	−2.763
Constant	13.2738***	0.0873833	151.9
MCMC monthly income before accidents happened	2.09157e−05**	8.60224e−06	2.431
Time	0.000776589***	0.000161590	4.806
Mental suffering	1.22153*	0.634478	1.925
Non-Cantonese speaker	−0.982315***	0.126674	−7.755

*** 0.001 level

8 Conclusions

Construction accidents lead to various losses such as grief and monetary loss. They also point to court hearings when there are disagreements concerning the amount of compensation. To study the impact of various factors which might affect the amounts of payments, we examined the court case reports for Hong Kong HKILL, from 2000 to 2015. Most of the victims had studied in high schools and were aged between 30 and 49.

As time series data often suffer from endogeneity, heteroscedasticity and serial correlation problems, we adopted the use of GMM-SYS because this can solve these problems; using this, we studied the various factors which affect the amount of accident compensation. It was found that there is no significant relationship between workers who have a mental disorder and the total amount of compensation. Court cases with more extensive hearings have a positive and significant relationship with the amount of compensation paid. Finally, non-Cantonese speakers receive less compensation.

As significant numbers of people from ethnic minorities have come to work in the construction industry, possibly a new area for research, stemming from this research, which would be valuable to pursue, would be related to how to improve safety standards in the Hong Kong construction industry for all of its workers, regardless of ethnic origin. For example, the following issues could be looked at whether the same safety standards are applied among non-Cantonese speakers as are applied among local people, and what are the significant factors that lead to the lower compensation levels among the former group of workers—especially considering that the Cap. 602 Race Discrimination Ordinance came into effect in July 2009 (Department of Justice 2016).

Acknowledgements This chapter is a revised version of Li, Rita Yi Man, Chau, Kwong Wing, Ho, Daniel Chi Wing (2017) Dynamic Panel Analysis on Construction accidents in Hong Kong, Asian Journal of Law and Economics, 8(3), 1–14.

References

Alanya, A., M. Swyngedouw, V. Vandezande, and K. Phalet. 2015, Fall. Close encounters: Minority and majority perceptions of discrimination and intergroup relations in Antwerp, Belgium. *International Migration Review* 1–25.
Bun, M. J. G., and F. Windmeijer. 2009. The weak instrument problem of the system GMM estimator in dynamic panel data models. *Econometrics Journal* 12: 1–32.
Department of Justice. 2016. *Cap. 602 race discrimination ordinance (current version)*. https://www.elegislation.gov.hk/index/chapternumber?QS_CAP_NO=RACIAL%20DISCRIMINATION%20ORDINANCE&p0=1&TYPE=1&TYPE=2&TYPE=3&LANGUAGE=E&TITLE=discrimination.

Feng, Y., and P. Wu. 2015. Risk compensation behaviours in construction workers' activities. *International Journal of Injury Control and Safety Promotion* 22: 40–47.

Garve, K. 1935. Physical pain and mental suffering in workmen's compensation cases. *Detroit Law Review* 5: 109–127.

Guryan, J., and K.K. Charles. 2013. Taste-based or statistical discrimination: The economics of discrimination reutrns to its roots. *The Economic Journal* 123: F417–F432.

Irumba, R. 2014. Spatial analysis of construction accidents in Kampala, Uganda. *Safety Science* 64: 109–120.

Labour Department. 2016. *Construction safety award scheme*. http://www.labour.gov.hk/eng/news/leafletE_CISAS_1617.pdf.

Leitão, N.C., and M. Shahbaz. 2013. Carbon dioxide emissions, urbanization and globalization: A dynamic panel data. *The Economic Research Guardian* 3: 22–32.

Li, R.Y.M. 2013. The usage of automation system in smart home to provide a sustainable indoor environment: A content analysis in web 1.0. *International Journal of Smart Home* 7: 47–60.

Li, R.Y.M. 2015a. *Construction Safety and Waste Management: An Economic Analysis*. Germany: Springer.

Li, R.Y.M., and S.W. Poon. 2009a. Future motivation in construction safety knowledge sharing by means of information technology in Hong Kong. *Journal of Applied Economic Sciences* 4: 457–472.

Li, R.Y.M., and S.W. Poon. 2009b. Workers' compensation for non-fatal construction accidents: Review of Hong Kong court cases. *Asian Social Science* 5: 15–24.

Li, R.Y.M., and Y.L. Li. 2013. The role of competition law: an Asian perspective. *Asian Social Science* 9: 47–53.

Li, R.Y.M., and S.W. Poon. 2013. *Construction safety*. Springer.

Li, R.Y.M., and H.Y. Tsoi. 2014. Latin America Sustainable Building Finance Knowledge Sharing. *Latin American Journal of Management for Sustainable Development* 1: 213–228.

Li, H., M. Lu, S.-C. Hsu, M. Gray, and T. Huang. 2015a. Proactive behavior-based safety management for construction safety improvement. *Safety Science* 75: 107–117.

Li, R. Y. M., A. Lee, A. Cheung, and H. Ma. 2015b. Unification of competition law in EU: Can that provide a useful example for Asian countries? In *EUSAAP annual conference*, Seoul.

Li, R.Y.M., H.C.Y. Li, C.K. Mak, T. Tang, and Beiqi. 2016. Sustainable smart home and home automation: Big data analytics approach. *International Journal of Smart Home* 10: 177–187.

Liao, C.W., and T.L. Chiang. 2015. The examination of workers' compensation for occupational fatalities in the construction industry. *Safety Science* 72: 363–370.

Mistikoglu, G., I.H. Gerek, E.P.E. Erdis, M. Usmen, H. Cakan, and E.E. Kazan. 2014. Decision tree analysis of construction fall accidents involving roofers. *Expert Systems with Applications* 42.

Occupational Safety and Health Branch Labour Department. 2016. Occupational safety and health statistics. *Bulletin*. http://www.labour.gov.hk/eng/osh/pdf/Bulletin2015_EN.pdf.

Panuwatwanich, K., S. Al-Haadir, and R.A. Stewart. 2017. Influence of safety motivation and climate on safety behaviour and outcomes: Evidence from the Saudi Arabian construction industry. *International Journal of Occupational Safety and Ergonomics* 23: 60–75.

Proehl, P.O. 1961. Anguish of mind damages for mental suffering under illinois law. *Northwestern University Law Review* 56: 477–501.

Quinlan, M., S.J. Fitzpatrick, L.R. Matthews, M. Ngo, and P. Bohle. 2015. Administering the cost of death: Organisational perspectives on workers' compensation and common law claims following traumatic death at work in Australia. *International Journal of Law and Psychiatry* 38: 8–17.

Rock, P.G., and P.C. Kitty. 2015. Twenty-first century environmental dispute resolution—Is there an 'ECT' in your future? *Journal of Energy and Natural Resources Law* 33: 10–33.

Roomkin, M., and R.N. Block. 1981. Case processing time and the outcome of representation elections: Some empirical evidence. *Univeristy of Illinois Law Review* 1981: 75–98.

Saracino, A., M. Curcuruto, G. Antonioni, M.G. Mariani, D. Guglielmi, and G. Spadoni. 2015. Proactivity-and-consequence-based safety incentive (PCBSI) developed with a fuzzy approach to reduce occupational accidents. *Safety Science* 79: 175–183.

Secretary for Sustainable Development. 2015. *WBTC No. 32/99 independent safety audit scheme for mega capital works contracts or capital works contracts involving unconventional construction method*. http://www.devb.gov.hk/filemanager/technicalcirculars/en/upload/184/1/wb3299.pdf.

Taleb, N.N., D.G. Goldstein, and M.W. Spitznagel. 2009. The six mistakes executives make in risk management. *Harvard Business Review* 1–4.

Wong, L., Y. Wang, P.E.T. Law, and C.T. Lo. 2016. Association of root causes in fatal fall-from-height construction accidents in Hong Kong. *Journal of Construction Engineering and Management* 142.

Yin, X., D. Levy, C. Willinger, A. Adouriand, and M.G. Larso. 2016. Multiple imputation and analysis for high-dimensional incomplete proteomics data. *Statistics in Medicine* 35: 1315–1326.

Zhao, X., and M. Biernat. 2017. "Welcome to the U.S." but "Change Your Name"? Adopting anglo names and discrimination. *Journal of Experimental Social Psychology* 70: 59–68.

Chapter 5
Near-Misses Reports in European and Asian Countries' Construction Industry

Abstract The term 'near-miss' refers to incidents that nearly lead to accidents but did not eventually. Studying near-miss reports is essential in preventing accidents; it is considered a necessary means to reduce the accidents and fatalities on the sites. In this chapter, we study near-miss reports in European and Asian countries.

Keywords Near-miss · Construction industry · Construction safety

1 Introduction

Near-misses represent hazardous situations which have nearly led to an accident, and if their causes are ignored in the construction process, an accident may well happen because of them (Gnoni and Lettera 2012; Jones et al. 1999). As near misses alert workers to the significance of safety practices on site and help construction workers to learn from their mistakes (Li and Poon 2009; Li 2015), reporting them can provide useful information for risk planning and minimise threats on site (Gnoni and Saleh 2017), some people advocate sharing information about near misses among workers (Li and Poon 2009). Near-miss reports can be found across many different industries such as the nuclear industry, oil mining, the chemical industry, construction and health care (Lander et al. 2011; Van der Schaaf 1995; Taylor and Lacovara 2015; Zeng et al. 2008). Learning lessons from near-miss reports is a well-known method to prevent more serious incidents (Raviv and Shapira 2018). Investigations of these events attempt to improve safety. They can be used to build a useful database that provides effective interventions for dealing with possible future failures. Recently, near misses have been increasingly used in the construction industry for risk identification (Cambraia et al. 2010; Cheng et al. 2012; Fabiano and Curro 2012; Lopez et al. 2008; Wright and Schaaf 2004).

Developing a matured near-miss reporting mechanism has apparent benefits at the industry level since the data can be extended to other industries in which similar situations may occur (Cambraia et al. 2010). The following sections are intended to show the near-miss reporting in construction safety, including the background, ideologies and methodologies used, the reports' applications in the industry, and their demand across some different countries.

© Springer Nature Singapore Pte Ltd. 2019
R. Y. M. Li, *Construction Safety Informatics*,
https://doi.org/10.1007/978-981-13-5761-9_5

2   The Rationale of Near-Miss Management

Near-miss reporting has been adopted in risk management in the construction sector (Gnoni and Saleh 2017; Jones et al. 1999). It usually involves stakeholders who are involved in all aspects of the production chain. In general, near-miss studies start from the individual level (learning from experience) and carry on up to the organisational, industry and regulatory levels (Hovden et al. 2011; Gnoni and Saleh 2017). There are different levels of near-miss studies concerning forming an integrated management system. Gnoni and Saleh (2017) suggested that a near-miss system aims to 'harvest value' from near-miss data by assessing and prioritising the risk implications, identifying the failure generating mechanisms, and guiding future interventions and safety improvements and awareness.

Historically, the European Union (EU) set up the European Commission's Joint Research Centre (ECJRC) in 1984, this organisation developed the Major Accident Reporting System (MARS) as a notification scheme mandated within the EU (also applicable to some non-EU countries such as Norway):

- to provide guidance concerning industrial accidents,
- to record and disseminate information,
- to provide an analysis of the causes of incidents, and
- to offer preventive measures (Seveso I 1982; Seveso II 1997; Seveso III 2012).

The near-miss reporting system of the EU (NMS) helps in evaluating the risk curve of some specific unexpected events (Kirchsteiger 2006). It was initiated by the Safety Management Systems (SMSs), while MARS (above) was one of the most effective systems at cross-national level (Gnoni et al. 2013; Gnoni and Lettera 2012; Gnoni and Saleh 2017). The NMS is concerned with how frequently and how quickly the causes of near-misses lead to an accident. It acts as an operational feedback process that attempts to collect data to perform analyses and make implications such as the need for safety interventions (Phimister et al. 2003).

Learning from near-misses is less costly than learning from accidents. The primary value of NMS is in the learning loop it provides within and across organisations—in focusing safety resources onto addressing unsafe acts, reducing hazardous conditions and procedures, and improving the design and operational safety issues (Gnoni and Saleh 2017). Particularly in construction safety management, the assessment of near-misses controls and monitors safety over time. Therefore, this current research is concerned with near misses that could help to create essential predictors which may notify of critical sequences of events. For instance, a failure in an emergency power system could sequentially lead to a core meltdown (Saleh et al. 2013).

In reality, the application of near-miss reporting mechanisms varies since an NMS may be constructed to deal with particular needs. Therefore, the later sections, here, about applications will show provisional applications in the field among different kinds of construction sites and in different countries.

3 Methods

It is commonly agreed that there can be no absolutely safe environment: errors can arise in every level in production; therefore, an NMS does not aim to eliminate the risk but to effectively reduce the potential risks through effective handling processes (Kessels-Habraken et al. 2010). Put straightforwardly, the methodology associated with near misses is that it is assumed that particular types of accidents have specific kinds of causes. Therefore, if one particular kind of near miss is noticed interventions can be applied to tackle the specific risk which causes the near miss—and so actual accidents can be avoided (Wright and Schaaf 2004).

However, due to the fact that construction processes are involved, they involve enormously complex causal patterns between near misses and accidents. Thus, one near-miss situation could be a causal predictor of a number of different consequences. For instance, based on the common cause hypothesis, the above condition shows the commonality of causes between different effects (Wright and Schaaf 2004). Graphically, the fact that a number of different near-miss states can trigger accident sequences is shown in Fig. 1 (Gnoni and Saleh 2017).

Although NMS is a broad system, it has specific overall objectives. According to Gnoni and Saleh (2017) and Van der Schaaf et al. (1991), the goals include to gain qualitative insights, to generate statistically reliable quantitative ideas and to sustain safety in operation. Phimister et al. (2003) introduced a brief design framework to

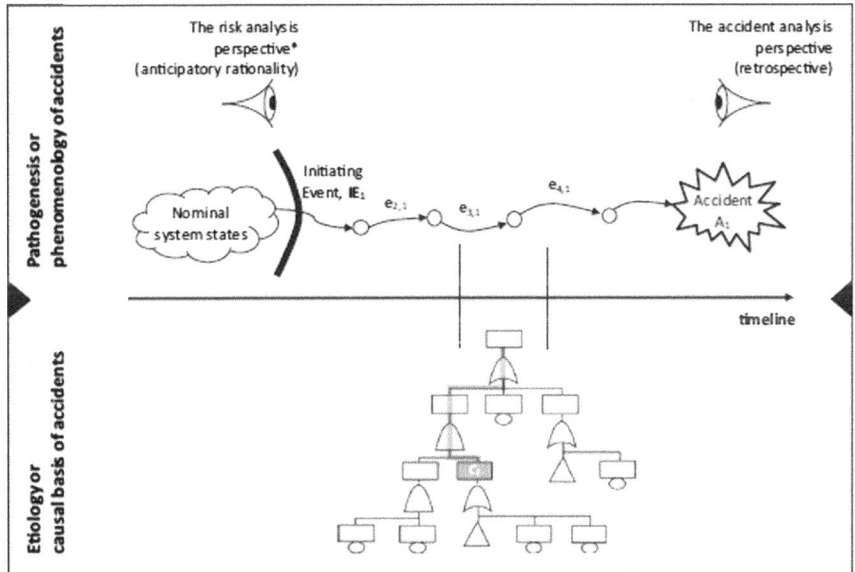

*Multiple accident trajectories are posited; only one is shown here for readability

Fig. 1 Accident sequences can be triggered by various near-miss conditions Gnoni and Saleh (2017)

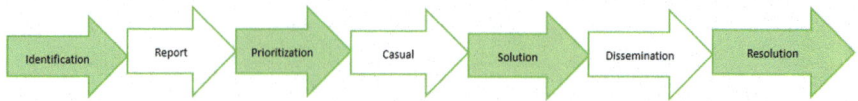

Fig. 2 Framework of how general knowledge is matched with actual needs (Gnoni and Saleh 2017; Phimister et al. 2003)

depict the design stages of an NMS which would achieve the objectives stated above. With these primary objectives in mind, an NMS has to be tailor-made to fit a specific industry, so that the general knowledge which has been acquired can eventually be matched with actual needs, as demonstrated in Fig. 2 (Muermann and Oktem 2002).

The usage of near-miss data, in actuality, has tended to emphasise the function of specific stages, such as identification, data analysis, or dissemination, among the tasks shown in the figure above (Cambraia et al. 2010). Later studies have criticised the efficiency of NMS as a system for enhancing construction safety, while some studies are concerned with particular stages of such a system (Cambraia et al. 2010).

Regarding the necessary context required for the study of near misses, some preconditions must be fulfilled before further analyses can take place. Those preconditions include an adequate separation between accidents and near misses, a classification of the possible injury severity, identification of the source of the hazardous situation, evaluation of the nature of the incidents' patterns and the establishment of an adequate methodology for dealing with the data collected (Raviv et al. 2017).

Typically, NMS begins with a qualitative (and occasionally quantitative) data collection process. In fact, at this stage, the raw data can be collected in many different forms such as documents, observations, measurements from apparatus and interviews. After obtaining the data, it is essential to establish a database which, for preference, will be based on a quantitative interpretation (Raviv et al. 2017). Using this database, different approaches are enabled in terms of the objective of analysis to find the causes of accidents, or, alternatively, of aiming to find out the specific situation which could possibly lead to accidents, and the sequential path wherein this situation exists. Based on different analyses or approaches, an NMS should be sufficient to notify or even provide suggestions for safety interventions to avoid risks, as long as there is enough data in the database to deal with a comprehensive set of situations. The above process and methodology can be graphically summarised in Fig. 3.

Although we have focused, here, on a brief explanation of the near-miss theory, in the literature, many studies focus on looking at the models which are in use. Some of these studies have effectively picked up the essential hypotheses in NMS modelling. Wright and Schaaf (2004) have indicated the difference between the ratio model and the common cause model, while the disparity in approaches can be summarised as resulting from the confounding of correlation and causality. Apart from this, there are some other approaches applicable to NMS modelling: the matrix approach and the proposed index approach (Gnoni and Lettera 2012). Although there are some

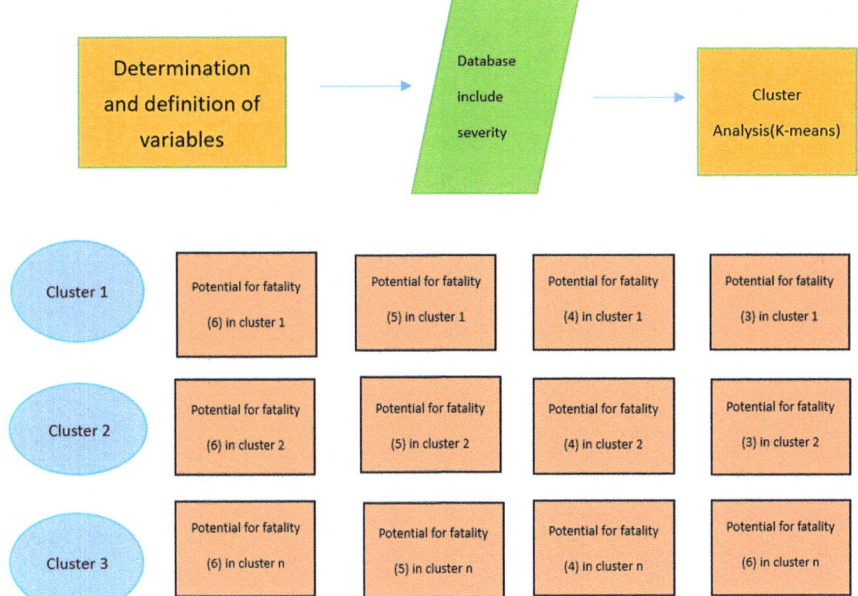

Fig. 3 The process and methodology of NMS. *Source* Raviv et al. (2017)

common models that can be found in the literature, NMS for construction safety or other specific areas have to be specifically designed prior to their use.

4 Application of Near-Miss Reporting Mechanisms in the Construction Industry Around the World

However, there is a scarcity of studies which can be found on the implementation of NMS to construction safety in Singapore. Besides, other models make use of ergonomics or geographic information system in modelling for construction safety study (Bansal 2011; Haslam et al. 2005).

There are some construction safety studies which do not look at applying NMS; however, there are studies in the literature which do examine the application of NMS. In order to investigate the railway operating system in the UK, Wright and Schaaf (2004) adopted a Confidential Incident Reporting and Analysis System (CIRAS). This work resulted in an important finding which consisted of a useful causal taxonomy of near-misses produced via a quantitative method. This taxonomy has helped the operational management of the UK railway network to record near-misses and handle the situation appropriately—according to the guide produced by CIRAS.

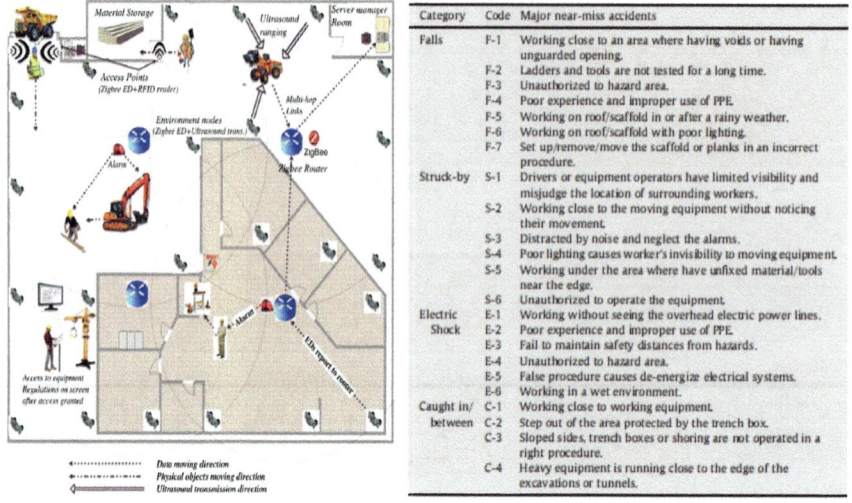

The table in the figure:

Category	Code	Major near-miss accidents
Falls	F-1	Working close to an area where having voids or having unguarded opening.
	F-2	Ladders and tools are not tested for a long time.
	F-3	Unauthorized to hazard area.
	F-4	Poor experience and improper use of PPE.
	F-5	Working on roof/scaffold in or after a rainy weather.
	F-6	Working on roof/scaffold with poor lighting.
	F-7	Set up/remove/move the scaffold or planks in an incorrect procedure.
Struck-by	S-1	Drivers or equipment operators have limited visibility and misjudge the location of surrounding workers.
	S-2	Working close to the moving equipment without noticing their movement.
	S-3	Distracted by noise and neglect the alarms.
	S-4	Poor lighting causes worker's invisibility to moving equipment.
	S-5	Working under the area where have unfixed material/tools near the edge.
	S-6	Unauthorized to operate the equipment.
Electric Shock	E-1	Working without seeing the overhead electric power lines.
	E-2	Poor experience and improper use of PPE.
	E-3	Fail to maintain safety distances from hazards.
	E-4	Unauthorized to hazard area.
	E-5	False procedure causes de-energize electrical systems.
	E-6	Working in a wet environment.
Caught in/ between	C-1	Working close to working equipment.
	C-2	Step out of the area protected by the trench box.
	C-3	Sloped sides, trench boxes or shoring are not operated in a right procedure.
	C-4	Heavy equipment is running close to the edge of the excavations or tunnels.

Fig. 4 The real-time tracking system of near-miss accidents (ARTTS-NMA). *Source* Wu et al. (2010b)

In construction site safety, early data collection has benefited from the Radio-Frequency Identification (RFID) and Wireless Sensor Network (WSN) technologies which have shifted NMS into the real-time tracking arena. This has meant that the responses to near-misses can be sped up—as measured in (Wu et al. 2010b; Yang et al. 2012). The latter paper describes the application of the above equipment to the real-time tracking of near-miss accidents in systems such as (ARTTS-NMA); this system can be used on construction site as is shown in Fig. 4. This system can be used to collect data regarding four main categories of near miss: falls, struck-by, electric shock and caught in/between.

Similarly, Wu et al. (2010a) developed another NMS which was focused on interrupting and preventing Precursors and Immediate Factors related to accidents (PaIFs) on construction sites. It applied a schematic model for the improvement of safety on sites, as the figure below shows. With regard to the real-time tracking of precursors and immediate factors, another systematic model has been applied to this. This framework provides near-miss reporting in order to enhance its model on a regular basis; also, real-time tracking improves the efficiency of accident preventive interventions. The mathematical expressions show the quantitative process carried out for data analysis in the model, i.e.

$$\text{Accident} = \text{Near Miss} + \text{Exacerbating factor(s)}$$
$$\text{Accident} = \text{Near Miss} - \text{Mitigating factor(s)}$$

These two equations indicate the occurrence of near-misses with either mitigating factors or exacerbating factor which would result in an actual accident. Therefore, (using this model) a general interpretation of the occurrence of an accident would be

$$\text{Accident} = \text{Near Miss} + \text{Exacerbating factor(s)} - \text{Mitigating factor(s)}$$

If the situation is not interrupted, a consequential accident in the future can be explained as

$$\text{Accident}_{t+1} = \text{Accident}_t + \emptyset + \text{time} - \text{mitigating factor(s)}$$

A graphical expression is shown in Fig. 5, which illustrates the mechanism of the model applied.

Similar methods can be found in other studies (e.g. Ale et al. 2008; Bellamy et al. 2007). There is a unique analytical tool named Storybuilder which analyses data held for the Netherland's construction industry to study the causes of accidents. In fact, other analytical tools, such as Storybuilder, have been designed to deal with a sequence of accidents such as the ones dealt with in the causal hypothesis model (Gnoni and Saleh 2017). However, the reporting processes for these do not use as much technology as does NMS (Wu et al. 2010a, b).

Cambraia et al. (2010) introduced the essential idea that focuses on the function and mission of different stages in design. This concerns actual practice in the identification of near misses: analysing and disseminating the estimated information (data could be translated from qualitative into quantitative data) processed by construction site safety reporting mechanisms. With the modifications of Saurin et al. (2004), Cambraia et al. (2010) adopted three hierarchical levels of decision-making with respect to time and uncertainty. It is crucial to point out that data collection depends to a very great extent on the willingness of workers on sites to participate in surveys.

5 Near-Miss Reporting Policies in the EU and Asia

Although the central concerns in terms of near-miss reporting are similar across geographical boundaries, the context varies a great deal among different countries, as accident management policies may be entirely different (Jones et al. 1999; Muermann and Oktem 2002). For instance, the level of technological advancement is generally lower among the developing countries than it is among the developed countries. A high-tech monitoring system could be ineffective in less developed regions due to there being insufficient knowledge about it and support.

Many studies have been conducted to establish the methodological framework of NMS (Gnoni et al. 2013; Gnoni and Lettera 2012; Gnoni and Saleh 2017; Jones et al. 1999; Meel et al. 2008; Wu et al. 2010a). Van der Schaaf (1995) suggested

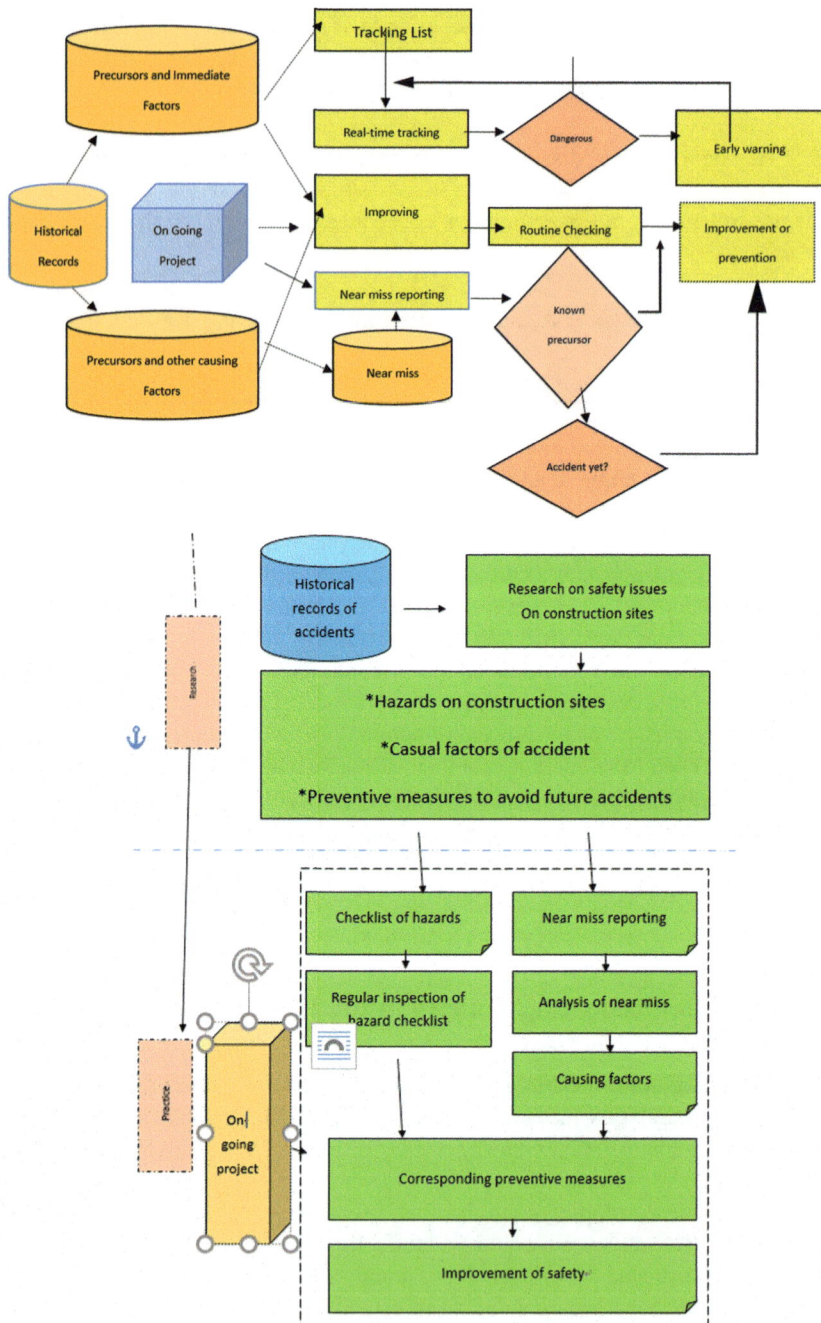

Fig. 5 Real-time tracking improves the efficiency of accident preventive intervention (Wu et al. 2010a)

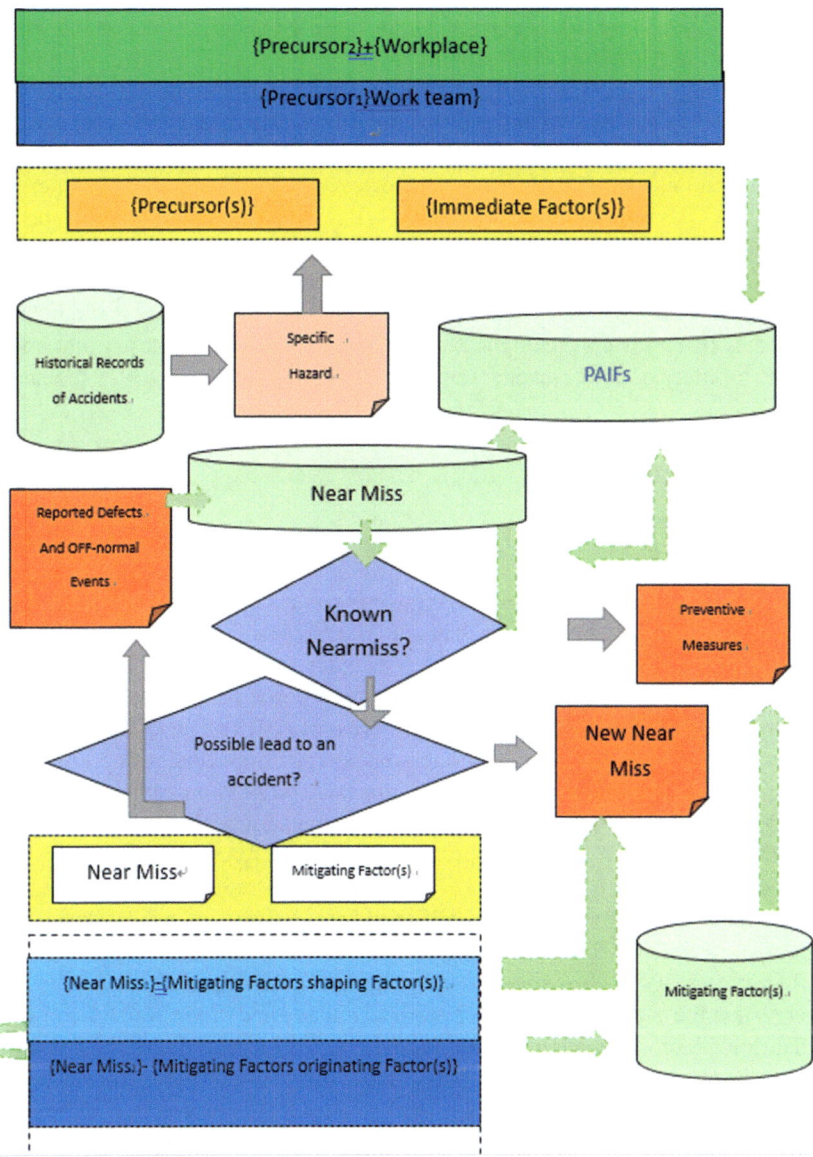

Fig. 5 (continued)

that three crucial stages are required for an active NMS: identification, analysis and dissemination (Cambraia et al. 2010).

Regarding the identification stage, hazardous situations can be directly harmful and even fatal, and they can be recorded in different ways: from obvious observation to accurate measurement. This implies that identification does not always require a complex technical monitoring system. However, it is undoubted that proper identification is a necessary condition for NMS (Phimister et al. 2003). After successful identification, the recorded data has to be analysed and integrated into a database. There are many approaches and models which can be used to analyse the data collected (Wu et al. 2010a). Based on a well-developed database, dissemination is the third step that transfers the implications of events for specific industry into improvements in safety policy or accident interventions (Sawacha et al. 1999; Teo and Ling 2006).

5.1   European Union

First of all, the European Union (EU) provides an excellent example of the application of a near-miss reporting mechanism. The EU has established the Major Accident Reporting System (eMARS) which includes an NMS that collaborating members and some external countries use to share information and knowledge with one another (Seveso I 1982; Seveso II 1997; Seveso III 2012). The EU's Joint Research Centre (JRE) attempts to exchange lessons learnt from accidents and near-misses about chemical hazard situations (The European Commission's Joint Research Centre 2012). The member countries have the responsibility to report any significant accident, as defined in Seveso III (2012), to the Major Accident and Hazards Bureau in the JRE. In fact, the application of the near-miss reporting mechanism here is remarkably successful, so that eMARS is able to evaluate the risk curve for specific types of accidental conditions (Gnoni and Lettera 2012; Kirchsteiger 2006). The effectiveness of eMARS is backed up by the compulsory participation of EU member so that the sources of information are adequate enough to allow for constructive analyses. Meanwhile, eMARS also accepts near-miss reporting in regard to other, unlisted, situations on a voluntary basis. Based on the implications from JRE, the EU drew up the Eurocodes, a set of European standards for construction, although near misses are not a significant concern in this guideline (The European Commission's Joint Research Centre 2016).

As well as eMARS, which is administered by JRE, the European Commission has another reporting mechanism to deal with offshore oil and gas incidents called Offshore Safety Directive Regulator (OSDR) (Health and Safety Execute 2015). The UK government has made it compulsory to report and investigate accidents and near-miss situations particular in regard to offshore oil and gas operations.

Apart from the chemical hazard situation, there are some NMSs which are applied individually in different countries which personally attempt to structuralize their near-miss reporting mechanisms. In the UK, there is a near-miss reporting scheme implemented by the Health and Safety Execute (HSE) (Health and Safety Execute 1998). The nature of accidents is defined separately under the Reporting of Injuries, Diseases and Dangerous Occurrences Regulations (RIDDOR) and Electricity Safety, Quality and Continuity Regulations (ESQCR); this is in order to handle different types of the accident report (Health and Safety Execute 2013). Some case studies have been looked in regard to a different sector in the UK, railways (Wright and Schaaf 2004). However, the NMS is voluntary: workers may report near misses if they wish to. An optional reporting mechanism is not as good as a compulsory NMS; however, the benefits in terms of safety improvements can still be observed.

The reporting mechanism issue has also been addressed in the United Kingdom. Despite the SOR system encouraging everyone on site to report unsafe acts or conditions, via either via computer or handwritten cards, for subsequent action by the health and safety team, problems with the SOR system emerged. These included the significantly increased administration efforts necessary to deliver predictable data; poor data quality; an unwelcome focus on the number rather than the content of the reports; their use as a tool to ascribe individual or organisational blame; and the perception that the SOR forms were being censored before they reached the health and safety team, which ultimately eroded trust between the workforce and management (Oswald et al. 2018). The authors warned that the system as implemented on the site in question had the potential to cause more harm than good, and both disengaged the workforce and frustrated the health and safety team. They further suggested that careful consideration of the process is required before implementation to avoid associations with blame, and the potential for a worker and H&S team disengagement from both the process of and health and safety management in practice (Oswald et al. 2018).

Besides the EU, there are some other countries which provide guidelines for near-miss reporting. The Organization for Economic Cooperation and Development (OECD) has imposed an NMS based on leading indicators concerning chemical incident (Cavalieri and Ghislandi 2008). With a historical approach, it constructs a reference model for the purposes of collecting target data and analysing them. Ultimately, it aims that the database will be accessible in relation to standard practices at the industrial level. It is also notable that, in the chemical sector, near-miss reporting is mandated and embedded into the Major Accident Hazard (MAH) legislation process (Phimister et al. 2003; Van der Schaaf 1995).

5.2 Near-Miss Reports in Asia

In India and some major developing countries, it is recorded that the prevalence of near-miss situations is relatively higher and, further, there is a higher mortality rate associated with them (Chhabra 2014). Therefore, the near-miss reporting primarily aims to evaluate the causes of mortality in order to make improvements in the relevant area. In China, the construction industry has applied the Occupational Health and Safety Assessment Series (OHSAS) 18000 system for NMS (Tam et al. 2004). This defines nine types of assessment as a reference for construction safety monitoring. However, it is also claimed that the support from the Chinese government is not sufficient to enable an effective near-miss reporting mechanism for the industry (Tam et al. 2004).

Similarly, in Thailand, awareness of construction safety is increasingly important, and so a great deal of research has been conducted to discover the factors which lead to accidents (Jitwasinkul et al. 2016). However, the concept of the near miss is under the radar, whilst the studies mentioned show the intention to identify near-miss situation; thus, there are various models applied in investigations, for instance, the Bayesian Belief Network (BBN) model is based on a probabilistic analytical approach and deals with a wide range of possible factors (Jitwasinkul et al. 2016). Therefore, the recent studies concerning safety in the construction industry in Thailand show that there is a developing awareness of how to create an active NMS while the government seems to lack the desire to provide guidance at the industrial level.

On the other hand, the Taiwan government have taken an active part in providing directions for the construction industry. It has adopted accident-mitigation strategies and put resources into general safety and health care. Meanwhile, it has been suggested that the (Taiwanese) government should provide further help and guidance to the industry on occupational safety issues (Cheng et al. 2012). Although the analysis process has been undertaken in the academic sector, the government has actively collected data from accident records to construct a database for the Council of Labor Affairs. With recommendations provided by the consultation bodies, it gives clear cause-and-effect analyses for the construction industry, through an NMS.

6 Future Development of Near-Miss Analysis

In terms of the future development and implementation of systems which deal with near misses, it is essential to enable the inclusive identification of potential risks across the entire range of severity outcome levels and to use a multi-criteria decision analysis in order to obtain the relative weights of the various severity outcome levels vis-à-vis the multiple factors (e.g. cost, reputation) that must be considered by the construction company when faced with the outcome of a safety incident (Raviv and Shapira 2018). In addition, a flexible safety management framework and a comprehensive safety monitoring system should be established. In developing this framework, the perceptions and attitudes of frontline safety supervisors and labourers towards the SMS should be explored, when such comprehensive information on the safety awareness and behaviour of these people are available, perhaps from an extensive survey in the future (Yiu et al. 2018). Finally, it has been argued that NMS is one of the pillars in the implementation of observability, in-depth, and that the boundaries with the two other components (fault detection/online monitoring, and inspection) are likely to be blurred in the future, and that the next generation of NMS will likely integrate data from multiple sources to improve the efficacy of precursor identification, prioritisation and of the safety interventions which result from the data, and ultimately of accident prevention. Further developments will be oriented towards validating the contributions of these safety principles through field testing (Gnoni and Saleh 2017).

References

Ale, B.J.M., L.J. Bellamy, H. Baksteen, M. Damen, L.H.J. Goossens, A.R. Hale, M. Mud, J. Oh, I.A. Papazoglou, and J.Y. Whiston. 2008. Accidents in the construction industry in the Netherlands: An analysis of accident reports using Storybuilder. *Reliability Engineering & System Safety* 93 (10): 1523–1533.

Bansal, V. K. 2011. Application of geographic information systems in construction safety planning. *International Journal of Project Management* 29:66–77.

Bellamy, L.J., B.J.M. Ale, T.A.W. Geyer, L.H.J. Goossens, A.R. Hale, J. Oh, M. Mud, A. Bloemhof, I.A. Papazoglou, and J.Y. Whiston. 2007. Storybuilder: A tool for the analysis of accident reports. *Reliability Engineering & System Safety* 92 (6): 735–744.

Cambraia, F. B., T. A. Saurin, and C. T. Formoso. 2010. Identification, analysis and dissemination of information on near misses: A case study in the construction industry. *Safety Science* 48 (1): 91–99.

Cavalieri, S., and W.M. Ghislandi. 2008. Understanding and using near-misses properties through a double-step conceptual structure. *Journal of Intelligent Manufacturing* 21 (2): 237–247.

Cheng, C.-W., S.-S. Leu, Y.-M. Cheng, T.-C. Wu, and C.-C. Lin. 2012. Applying data mining techniques to explore factors contributing to occupational injuries in Taiwan's construction industry. *Accident Analysis and Prevention* 48: 214–222.

Chhabra, P. 2014. Maternal near miss: An indicator for maternal health and maternal care. *Indian Journal of Community Medicine* 39 (3): 132–137.

Fabiano, B., and F. Curro. 2012. From a survey on accidents in the downstream oil industry to the development of a detailed near-miss reporting system. *Process Safety and Environmental Protection* 90 (5): 357–367.

Gnoni, M. G., S. Andriulo, G. Maggio, and P. Nardone. 2013. "Lean occupational" safety: An application for a near-miss management system design. *Safety Science* 53: 96–104.

Gnoni, M.G., and G. Lettera. 2012. Near-miss management systems: A methodological comparison. *Journal of Loss Prevention in the Process Industries* 25 (3): 609–616.

Gnoni, M.G., and J.H. Saleh. 2017. Near-miss management systems and observability-in-depth: Handling safety incidents and accidents precursors in light of safety principles. *Safety Science* 91: 154–167.

Haslam, R.A., S.A. Hide, A.G.F. Gibb, D.E. Gyi, T. Pavitt, S. Atkinson, and A.R. Duff. 2005. Contributing factors in construction accidents. *Applied Ergonomics* 36 (4): 401–415.

Health and Safety Execute. 1998. *MHIDAS-major hazards incident data system* [cited 06/10/2017. Available from http://www.hse.gov.uk/research/journals/mrn698a.htm.

Health and Safety Execute. 2013. *RIDDOR-reporting of injuries, diseases and dangerous occurrences regulations* [cited 06/11/2017. Available from http://www.hse.gov.uk/riddor/index.htm.

Health and Safety Execute. 2015. *Offshore safety directive regulator (OSDR)* [cited 06/09/2017. Available from http://www.hse.gov.uk/osdr/index.htm.

Hovden, J., F. Stroseth, and R.K. Tinmannsvik. 2011. Multilevel learning from accidents: Case studies in transport. *Safety Science* 49 (1): 98–105.

Jitwasinkul, B., B.H.W. Hadikusumo, and A.Q. Memom. 2016. A Bayesian belief network model of organizational factors for improving safe work behaviors in Thai construction industry. *Safety Science* 82: 264–273.

Jones, S., C. Kirchsteiger, and W. Bjerke. 1999. The importance of near miss reporting to further improve safety performance. *Journal of Loss Prevention in the Process Industries* 12 (1): 59–67.

Kessels-Habraken, M., T. Van der Schaaf, J.D. Jonge, and C. Rutte. 2010. Defining near misses: Towards a sharpened definition based on empirical data about error handling processes. *Social Science & Medicine* 70 (9): 1301–1308.

Kirchsteiger, C. 2006. Impact of accident precursors on risk estimates from accident database. *Journal of Loss Prevention in the Process Industries* 19: 630–638.

Lander, L., E. Eisen, T. Stentz, S. Kathleen, B. Wendland, and M. Perry. 2011. Near-miss reporting system as an occupational injury preventive intervention in manufacturing. *American Journal of Industrial Medicine* 54 (1): 40–48.

Li, R.Y.M. 2015. *Construction safety and waste management: An economic analysis*. Switzerland: Springer.

Li, R.Y.M., and S.W. Poon. 2009. Future motivation in construction safety knowledge sharing by means of information technology in Hong Kong. *Journal of Applied Economic Sciences* IV 3 (9): 457–472.

Lopez, M.A., D.O. Ritzel, I. Fontaneda, and G. Alcantara. 2008. Construction industry accidents in Spain. *Journal of Safety Research* 39 (5): 497–507.

Meel, A., S. Warren, and U.G. Oktem. 2008. Analysis of management actions, human behavior, and process reliability in chemical plants. II. Near-miss management system selection. *Process Safety Progress* 27 (2): 139–144.

Muermann, A., and U.G. Oktem. 2002. The near-miss management of operational risk. *The Journal of Risk and Finance* 4 (1): 25–36.

Oswald, D., F. Sherratt, and S. Smith. 2018. Problems with safety observation reporting: A construction industry case study. *Safety Science* 107: 35–45.

Phimister, J.R., U.G. Oktem, P. Kleindorfer, and H. Kunreuther. 2003. Near-miss incident management in the chemical process industry. *Risk Analysis* 23 (3): 445–459.

Raviv, G., and A. Shapira. 2018. Systematic approach to crane-related near-miss analysis in the construction industry. *International Journal of Construction Management* 18 (4): 310–320.

Raviv, G., B. Fishbain, and A. Shapira. 2017. Analyzing risk factors in crane-related near-miss and accident reports. *Safety Science* 91: 192–205.

Saleh, J.H., E.A. Saltmarsh, F.M. Favaro, and L. Brevault. 2013. Accident precursors, near-misses, and warning signs: Critical review and formal definitions within the framework of discrete event systems. *Reliability Engineering & System Safety* 114: 148–154.

Saurin, T.A., C.T. Formoso, and L.B.M. Guimaraes. 2004. Safety and production: An integrated planning and control model. *Construction Management and Economics* 22 (2): 159–169.

Sawacha, E., S. Naoum, and D. Fong. 1999. Factors affecting safety performance on construction sites. *International Journal of Project Management* 17 (5): 309–315.

Seveso I. 1982. Council directive 82/501/EEC on the major-accident hazards of certain industrial activities. *Official Journal of the European Communities*. (Luxembourg).

Seveso II. 1997. Council directive 96/82/EC of 9 December 1996 on the control of major-accident hazards involving dangerous substances. *Official Journal of the European Communities*. (Luxembourg).

Seveso III. 2012. Council directive 2012/18/EU of the European Parliament and of the Council of 4 July 2012 on the control of major-accident hazards involving dangerous substances, amending and subsequently repealing Council Directive 96/82/EC Text with EEA relevance. *Official Journal of the European Communities*. (Luxembourg).

Tam, C.M., S.X. Zeng, and Z.M. Deng. 2004. Identifying elements of poor construction safety management in China. *Safety Science* 42 (7): 569–586.

Taylor, J.A., and A.V. Lacovara. 2015. From infancy to adolescence: the development and future of the national firefighter near-miss reporting system. *New Solutions* 24 (4): 555–576.

Teo, E.A.L., and F.Y.Y. Ling. 2006. Developing a model to measure the effectiveness of safety management systems of construction sites. *Building and Environment* 41 (11): 1584–1592.

The European Commission's Joint Research Centre. 2012. *EMARS-major accident reporting system*, 09/06/2012 [cited 06/10/2017. Available from https://emars.jrc.ec.europa.eu/.

The European Commission's Joint Research Centre. 2016. *Standards in construction: The Eurocodes*, 07/14/2016 [cited 06/10/2017. Available from https://ec.europa.eu/jrc/en/research-topic/standards-construction-eurocodes.

Van der Schaaf, T. 1995. Near miss reporting in the chemical process industry: An overview. *Microelectronics Reliability* 35 (9–10): 1233–1243.

Van der Schaaf, T., D.A. Lucas, and A.R. Hale. 1991. *Near-miss reporting as a safety tool*. Oxford: Butterworth-Heinemann.

Wright, L., and T. Van der Schaaf. 2004. Accident versus near miss causation: a critical review of the literature, an empirical test in the UK railway, and their implications for other sectors. *Journal of Hazardous Materials* 111 (1–3): 105–110.

Wu, W., A.G.F. Gibb, and Q. Li. 2010a. Accident precursors and near misses on construction sites: An investigative tool to derive information from accident databases. *Safety Science* 48 (7): 845–858.

Wu, W., H. Yang, D.A.S. Chew, S. Yang, A.G.F. Gibb, and Q. Li. 2010b. Towards an autonomous real-time tracking system of near miss accidents on construction sites. *Automation in Construction* 19 (2): 134–141.

Yang, H., D.A.S. Chew, W. Wu, and Q. Li. 2012. Design and implementation of an identification system in construction site safety for proactive accident prevention. *Accident Analysis and Prevention* 48: 193–203.

Yiu, N.S., N. Sze, and D.W. Chan. 2018. Implementation of safety management systems in Hong Kong construction industry—A safety practitioner's perspective. *Journal of safety research* 64: 1–9.

Zeng, S.X., V.W.Y. Tam, and C.M. Tam. 2008. Towards occupational health and safety systems in the construction industry of China. *Safety Science* 46 (8): 1155–1168.

Chapter 6
Construction Safety Knowledge Sharing: A Psychological Perspective

Abstract The recent popularity of Web 2.0 mobile apps and the Internet of Things provides a new perspective on asynchronous safety knowledge sharing. Workers from different teams who never meet can share knowledge easily by straightforward mechanisms. A natural starting point when conceptualising the role of economic institutions is to consider the incentives which affect people's activities. This also matches the psychological theory well as that studies the factors that impact our behaviour. In this research paper, we examine the construction practitioners' knowledge-sharing behaviour from (1) new institutional economic perspectives under the lens of informal institutions and (2) psychology's perspectives, such as Homan's proposition.

Keywords Construction safety · Knowledge sharing · Psychology · New institutional economics

1 Introduction

The construction industry is infamous for the large numbers of accidents which are associated with it (Li and Poon 2011; Li 2018). Just looking at China, there was an average of more than 2500 accident deaths annually in the construction industry from 1997 to 2014. The construction industry is also recognised as treacherous even in developed countries such as the US and the UK. Global statistical data has recorded the accidental death and injury rate in the construction industry as being two to three times higher than the average in other industries. In spite of more attention being paid to safety management recently, the accident rate continues to be high (Guo et al. 2017). The industry suffers the largest number of fatal accidents among all occupations and it is only second to the manufacturing industry regarding serious accidents, according to Eurostat (2016) (Fig. 1).

A construction site can be a very complicated and intricate environment, accommodating more than 20 trades with many different skilled and unskilled workers involved in various construction processes. Furthermore, as different teams may work at a different time on sites, the transmission of safety knowledge from main

© Springer Nature Singapore Pte Ltd. 2019
R. Y. M. Li, *Construction Safety Informatics*,
https://doi.org/10.1007/978-981-13-5761-9_6

contractors to sub-sub-sub-contractors, or even to subcontractors at lower levels, may be difficult due to the presence of asynchronous knowledge sharing. This situation leads to construction being a high safety risk industry (Le et al. 2014).

The recent rise in the use of the Internet of Things, Web 2.0, etc. may offer some useful mechanisms whereby safety knowledge may be shared. Many Web 2.0 tools have entered into widespread use. For example, Twitter had 320 million users as of March 2016 and Facebook had 1.59 billion monthly active users as of December 2015. Furthermore, 70% of Facebook users interact with the platform daily, compared with 38% of Twitter users (Kim and Hastak 2018).

These dynamic Web 2.0 applications allow users to change content on the Internet (Li and Poon 2011) and offer an optimum solution to all knowledge seekers. Dave and Koskela (2009) developed an online forum designed to test and analyse participation in construction knowledge sharing. Unlike the traditional means of communication, such as phone calls and emails, social media systems often present different types of information from different contributors within a single message pane; this juxtaposition of potential influences challenges traditional health communication processing (Walther et al. 2018).

Besides, many different mobile apps can be used to share construction safety knowledge. For example, Red Cross First Aid includes information about crucial first-responder skills in their app library, 'American Red Cross'. The TrueLook apps enable users to utilise drones to take aerial photographs of job sites and the photos can then be organised in the app by the project (TSheets 2018). Concerning a different requirement, an ontology-based NoSQL database, Cassandra, can be used for supply chain data distribution (Moumita et al. 2015).

Similar to Web 2.0, the Internet of Things (IoT) can also be used for knowledge sharing. It integrates smart nodes, sensors and objects which can communicate with each other in the absence of human intervention (Conti et al. 2018). It enables knowledge exchange via a network of many connected 'things'. IoT can be applied on site to alert the construction workers with regards to safety risks and hazards. For example, when the weather is very hot and the workers have worked for more than 3 hours, a voice message can be sent to the workers to remind them to drink water (Li 2017).

IoT can also be identified by, for instance, QR codes and Radio-Frequency Identification (RFID) (Zhao et al. 2016). The ease with which dynamic QR codes can be changed, in terms of their interpretation, alters the face of knowledge sharing. For example, we can easily convert the abovementioned Facebook group's website to use dynamic QR codes, i.e. the content represented by the QR code can be changed by simply changing the website content (Fig. 2). Users can easily access such dynamic knowledge by scanning the dynamic QR code to discover useful and timely information (Fig. 3). Shin et al. (2011) proposed an RFID-based framework that integrates legacy systems with websites to facilitate communication, but knowledge sharing does not depend only on the existence of technologies and ontologies. The willingness to use technologies for sharing construction safety knowledge is also crucial. Nourbakhsh et al. (2012) set up a system which used the Microsoft Office SharePoint Server mobile app to report on-site accidents in Malaysia.

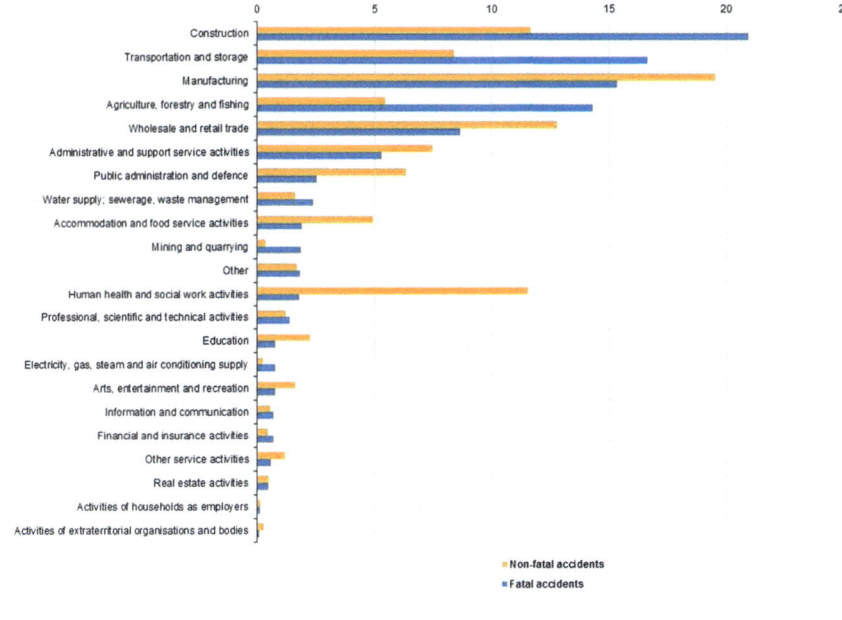

Note. Provisional.

Fig. 1 Serious and fatal accidents at work (Eurostat 2016)

Fig. 2 Construction safety group on Facebook (https://www.facebook.com/Construction-Safety-1390450597880578/)

Fig. 3 Dynamic QR code

2 A Psychological Perspective on Construction Safety

Psychology is one of the important factors that affect human's decision-making. Construction safety, which is affected by human acts and behaviour, is of no exception. Powell and Dalton (2003) suggested that a person is more likely to repeat an action when one or both of the following propositions are valid:

(1) The value proposition, the more valuable the consequence of an action;
(2) The success proposition, the more often the subject has been rewarded for the specific action.

Nevertheless, little research has been carried out on the ramifications of Homans' recommendations on knowledge sharing in the construction industry. Wang et al. (2014) suggested that the properties of shared knowledge, expected rewards and incentives affected the likelihood of knowledge sharing in high-technology firms. However, research into any of these psychological factors is scarce.

Leader–Membership Exchange theory concerns the vertical connections among administrators and subordinates (Krafft et al. 2012). It states that the nature of the dyadic relationship among leaders and their followers influences subordinates' states of mind and their conduct (Kraft et al. 2018). Top notch Leader–Membership Exchange prompts resources being available to subordinates, including knowledge which is not open to all (Krafft et al. 2012). Previous research has found that employees of a cell phone company with better Leader–Membership Exchange, demonstrated by a high degree of common regard, commitment and trust have been found to have a higher likelihood of sharing their insights and thoughts (Su et al. 2013).

Perceived Organisation Support under the social exchange theory may affect safety performance at work (Li 2015). It represents how much an organisation is concerned about and values their employees' commitment to the company and their employees' well-being, prompting the desire that their endeavours will be compensated. Higher Perceived Organisation Support is related to there being a more pronounced feeling of responsibility to the organisation among the staff (van Knippenberg et al. 2015). Previous research has demonstrated that better Perceived Organisation Support has had a hugely beneficial impact on the level of knowledge sharing in the Norwegian oil and gas industry (Nesheim and Smith 2015).

3 Informal Institution Settings and Knowledge Sharing

Traditional new institutional economists consider that there are formal and informal institutions. The formal institution, in general, refers to the black and white laws which include legislation and codes. The informal institution, i.e. any rules that are not in black and white such as culture, affects people's behaviour in various settings (Li and Chau 2016; Li 2012; Li and Poon 2009). Previous research has shown that there are different perceptions among construction practitioners about emails and in-person discussion. The chapter revealed that there is no difference between generations regarding the amount of knowledge sharing that takes place through in-person discussions and the amount which takes place via email, however, there are significant differences between generations regarding the use of instant messaging as opposed to meetings.

Given this, we conjecture that workers in various age and ethnic groups may have differing levels of eagerness to share knowledge using the Internet of Things, Web 2.0 and mobile applications because of various cultural and social standards.

4 Conclusion

The fact that the conditions which prevail on construction sites are always changing implies that having an appropriate knowledge-sharing platform is important. From this point of view, mobile apps, the Internet of Things (IoT) and Web 2.0 provide promising answers to the problems of sharing asynchronous knowledge on sites—problems which result from differences in the working times amongst different trades and high-level subcontractors.

Numerous studies concerning safety knowledge-sharing technologies have investigated the development of new, groundbreaking tools, but not the willingness to utilise these, which may depend on institutional differences emerging from age or ethnicity related or cultural differences or the psychological factors identified in Homan's propositions: Perceived Organisation Support and Leader Membership Exchange.

Acknowledgements An earlier version of the chapter appears in Li, Rita Yi Man, Kwong Wing Chau, Daniel Chi Wing Ho, Weisheng Lu and Mandy Wai Yee Lam (2018) Construction safety knowledge sharing by Internet of Things, Web 2.0 and mobile apps: psychological and new institutional economics, IOP Science Conference Series: Materials Science and Engineering (MSE) (Scopus), XXI International Scientific Conference Construction the Formation of Living Environment, Moscow, Russia, IOP Conference Series: Materials Science and Engineering, Volume 365, pp. 1–8.

References

Conti, Mauro, Ali Dehghantanha, K. Franke, and S. Watson. 2018. Internet of Things security and forensics: Challenges and opportunities. *Future Generation Computer Systems* 78: 544–546.

Dave, B., and L. Koskela. 2009. Collaborative knowledge management—A construction case study. *Automation in Construction* 18 (7): 894–902.

Eurostat. 2016. *Euro stat statistics explained* [cited 24 Jan 2018]. Available from http://ec.europa.eu/eurostat/statistics-explained/index.php/Main_Page.

Guo, H., Y. Yu, and M. Skitmore. 2017. Visualization technology-based construction safety management: A review. *Automation in Construction* 73: 135–144.

Kim, J., and M. Hastak. 2018. Social network analysis: Characteristics of online social networks after a disaster. *International Journal of Information Management* 38 (1): 86–96.

Krafft, M., T.E. DeCarlo, F.J. Poujol, and J.F. Tanner. 2012. Compensation and control systems: A new application of vertical dyad linkage theory. *Journal of Personal Selling & Sales Management* 32 (1): 107–115.

Kraft, A., J.L. Sparr, and C. Peus. 2018. Giving and making sense about change: The back and forth between leaders and employees. *Journal of Business and Psychology* 33 (1): 71–87.

Le, Q.T., D.Y. Lee, and C.S. Park. 2014. A social network system for sharing construction safety and health knowledge. *Automation in Construction* 46: 30–37.

Li, R.Y.M. 2012. Econometric modelling of risk adverse behaviours of entrepreneurs in the provision of house fittings in China. *Construction Economics and Building* 12 (1): 11.

Li, R.Y.M. 2015. *Construction safety and waste management: An economic analysis.* Switzerland: Springer.

Li, R.Y.M. 2017. Smart construction safety in road repairing works. *Procedia Computer Science* 111: 301–307.

Li, R.Y.M. 2018. *An economic analysis on automated construction safety: Internet of Things, artificial intelligence and 3D printing.* Singapore: Springer.

Li, R.Y.M., and K.W. Chau. 2016. *Econometric analyses of international housing markets.* Routledge.

Li, R.Y.M., and S.W. Poon. 2009. Future motivation in construction safety knowledge sharing by means of information technology in Hong Kong. *Journal of Applied Economic Sciences* IV (3(9)): 457–472.

Li, R.Y.M., and S.W. Poon. 2011. Using Web 2.0 to share the knowledge of construction safety as a public good in nature among researchers: the fable of economic animals. *Economic Affair* 31 (1): 73–79.

Moumita, D., J.C.P. Chen, and K.H. Law. 2015. An ontology-based web service framework for construction supply chain collaboration and management. *Engineering, Construction and Architectural Management* 22 (5): 551–572.

Nesheim, T., and J. Smith. 2015. Knowledge sharing in projects: Does employment arrangement matter? *Personnel Review* 44 (2): 255–269.

Nourbakhsh, M., Z.R. Mohamad, J. Irizarry, S. Zolfagharian, and M. Gheisari. 2012. Mobile application prototype for on-site information management in construction industry. *Engineering, Construction and Architectural Management* 19 (5): 474–494.

Powell, K.H., and M.M. Dalton. 2003. Co-production, service exchange networks, and social capital. *The Social Policy Journal* 2 (2–3): 89–106.

Shin, T.-H., S. Chin, S.-W. Yoon, and S.-W. Kwon. 2011. A service-oriented integrated information framework for RFID/WSN-based intelligent construction supply chain management. *Automation in Construction* 20 (6): 706–715.

Su, T., Z. Wang, X. Lei, and T. Ye. 2013. Interaction between Chinese employees' traditionality and leader-member exchange in relation to knowledge-sharing behaviors. *Social Behavior and Personality: An International Journal* 41: 1071–1082.

TSheets. 2018. *Top 10 construction apps for 2018* [cited 4 March 2018]. Available from https://www.tsheets.com/best-construction-apps.

van Knippenberg, D., J.-W. van Prooijen, and E. Sleebos. 2015. Beyond social exchange: Collectivism's moderating role in the relationship between perceived organizational support and organizational citizenship behaviour. *European Journal of Work and Organizational Psychology* 24 (1): 152–160.

Walther, J.B., J.-W. Jang, and A.A. Hanna Edwards. 2018. juxtaposition. *Health Communication* 33 (1): 57–67.

Zhao, C., J. Liu, F. Shen, and Y. Yi. 2016. Low power CMOS power amplifier design for RFID and the Internet of Things. *Computers & Electrical Engineering* 52: 157–170.

Chapter 7
Modelling the Construction Accident Cases via Structural Equation Modelling

Abstract Construction accident compensation is one of the most critical factors that motivate contractors and clients to provide sufficient safety measures for workers. A court will usually collect information about the average monthly earnings of the workers before the accident to evaluate their working ability and consider the effects of injury on their productivity; this is to provide a fair judgment relating to the compensation items. Regarding the level of compensation, most compensation categories such as PSLA, pretrial loss of earning and future treatment are reflected in the level of compensation. The more severe the injury caused by accident, the more is the compensation for the future loss. Besides, even though there are few studies concerning the economic aspects of tort, there is a distinct lack of studies on the factors that affect the compensation. This chapter fills this academic void by utilising a structural equation modelling approach. Accident compensation court cases dated from 1982 to 2015 in Hong Kong were collected. 14 categories of information (including the level of injury as a 'latent' or unmeasurable variable) were recorded and used to construct the model which was then used for compensation estimation.

1 Introduction

Safety is an essential issue in the construction industry. Guo et al. (2017) have suggested that construction safety issues have become important in research and practice in recent years mostly because of the high accident and death rates in the construction industry. Misiurek and Misiurek (2017) indicated that the average accident level was much higher in the construction industry than in the manufacturing industry—in the UK. Bunting et al. (2017) pointed out that, in the US, falls were the leading cause of death and the third leading cause of non-fatal injuries in the construction industry. Chen et al. (2017) suggested that a good climate not only affected construction workers' safety performance but also, indirectly, their levels of psychological stress. Besides, previous research has indicated that accidents are often caused by failures of technology, people or a combination of both. Indeed, the causes seldom consist of one simple issue but are complex constellations comprising system properties, events and existing preconditions (Li et al. 2017a). Li et al. (2017b) surveyed

© Springer Nature Singapore Pte Ltd. 2019

R. Y. M. Li, *Construction Safety Informatics*,

https://doi.org/10.1007/978-981-13-5761-9_7

Nanjing, China, and also found that the climate factor was critical in construction safety. Also, they also pointed out that safety management involvement and safety-related personnel support were important for workers' safety.

Since construction accidents often lead to grief and insurmountable monetary expenditure, construction industry practitioners and academic researchers provide ample suggestions regarding improving site safety. For example, Li and Poon (2011) pointed out that web 2.0 allows people to communicate efficiently and to make improvements relating to construction safety issues. Liu and Hu (2017) believed that information technology helped to reduce construction accidents by providing active safety control regarding the machines in use. Mohammad and Hadikusumo (2017) realised that technology could also be used for multilevel interventions during the construction process to reduce the level of accident risk. Azhar (2017) suggested that visualisation technologies (such as Building Information Modelling (BIM), four-dimensional simulations and three-dimensional immersive virtual reality environment) could be used by contractors, architects and engineers to virtually access the construction site's conditions and identify the possible risk which might emerge during the construction process.

Many experts confirm that there is a direct relationship between the number and severity of construction accidents and the compensation payable. Li (2015) indicated that construction accident rates are high in many places and this leads to higher compensation, loss in human resources and extensions to construction times. Park and Bhattacharya (2013) indicated that work-related injury or illness could cause additional costs that might not be covered by the workers' compensation payments. This led to a discussion of the effects of continuing impairment (early termination or prolonging) on employment. Li and Poon (2009) undertook a comprehensive study of non-fatal accident compensation court cases from 2004 to 2008 in Hong Kong and found that most victims were compensated under loss of earnings. Liao and Chiang (2015) researched the Taiwanese construction industry and found that workers involved in accidents while employed on public projects did not receive enough compensation and the amount of compensation paid varied according to the owners of the site concerned. Torres and Jain (2017) compared the non-economic damage (such as physical injury, inconvenience, loss of self-confidence, loss of esteem, mental distress, etc.) as related to the compensation levels in Chile and England. They found that England had made significant progress towards providing compensation for non-economic damage, while Chile still had lots of challenges and needed to improve in this area.

2 Research Method

Data was collected from 320 court cases dated from 1982 to 2015, which were all related to construction accidents in Hong Kong. 14 categories of information were recorded and used to construct the model, and some data have been estimated. These 14 variables were the age of plaintiff, the gender of the plaintiff, the plaintiff's

Table 1 Details of variables (Glofcheski 2012)

Name of the variables	Details
Pretrial loss of earning	Financial losses incurred as a result of the accident such as medical expenses and loss of income
Special damage	A kind of pretrial loss compensation award related to financial loss such as expenses for wheelchair and crutches or compensation for family members acting as carers for the plaintiff
Future loss of earning	A component of the injury reward which is calculated according to the income the plaintiff received before the accident. The calculation takes the expected future disability and the extent of disability due to the accident into account
Loss of earning capacity	Another compensation award is given to the plaintiff based on the probability that they might lose their job at some time in the future because of the injury suffered
Pain, suffering, loss of amenities (PSLA)	A kind of compensation award related to non-financial damage. It considers the physical pain and harm to life due to the injury. 'Loss of Amenities' refers to the situation where the plaintiffs' ability to enjoy their life is diminished because of the accident; this can be compensated for by the award of PSLA. The pre-accident life of the plaintiff will be examined to formulate the compensation amount
Future treatment	Future treatment refers to the medical expenses expected to be used for recovery in the future

monthly income before the accident, their future loss of earnings, their pretrial loss of earnings, their loss of earning capacity, any particular damage, future treatment, the death of the plaintiff, the duration of judgment, the level of court, contributory negligence and PSLA. Some of these items are further explained in Table 1. Finally, a latent variable called the level of injuries (Inj) has also been included in the model.

2.1 Structural Equation Modelling (SEM) Approach

The structural equation modelling approach is a favourite analysis technique because of the full range of questions that it can help answer (Schreiber 2008). Wang and Wang (2012) believed that SEM could be used to estimate a cluster of dependent variables without an upper limit in the model, which provides a flexible ability to construct the model. This approach can test over the fitness of the model and can estimate the direct and indirect effects across the variables. SEM can also handle complicated data like count and inevitable outcomes, non-normal outcome, censored outcome, etc. The function of the general structural equation is separated into two parts: the structural part linking latent variables to each other via systems of simultaneous equations and the measurement part, which connects latent variables to observed variables via a restricted (confirmatory) factor model (Jöreskog 1973).

2.1.1 Model Formation

SEM assesses structural and measurement model. A measurement model examines the relationships between the latent factors and observed variables. A confirmatory factor analysis tests the predetermined structure of the relationships among the variables in a model (Chen et al. 2018). Angelini and Heuvelink (2018) suggested that the system of equations of SEM was characterised by the structural model and the measurement model.

The structural model can be defined as

$$\eta = B\eta + \Gamma\xi + \zeta$$

$$E[\xi] = E[\zeta] = 0$$

$$\mathrm{Var}(\xi) = \Phi, \ \mathrm{Var}(\zeta) = \Psi$$

where η is a vector of latent endogenous variables (i.e. dependent variables), ξ a vector of latent exogenous variables (i.e. independent variables) and ζ a vector of structural errors. The diagonal elements of B are zero, ξ and ζ are mutually independent and normally distributed. Their variance–covariance matrices are given by Φ and Ψ, respectively.

The measurement model can be shown as

$$Y = K\eta + \varepsilon$$

$$X = \Lambda\xi + \delta$$

$$E[\epsilon] = E[\delta] = 0$$

$$\mathrm{Var}(\varepsilon) = \Theta_\epsilon, \ \mathrm{Var}(\delta) = \Theta_\delta$$

where ε and δ are mutually independent customarily distributed variables that are independent of all previously defined variables. The author assumed that K and Λ are identity matrices (hence $p = n$ and $q = m$). Note also that all variables are considered to have zero mean.

The SEM framework could also be shown as

$$y_i = \mu + \wedge \omega_i + \epsilon_i, \quad i = 1, \ldots, n,$$

where $y_i = (y_{i1}, \ldots, y_{ip})^{\mathrm{T}}$ is the observed random vector of subject i, $\epsilon_i = N_p(0, \Psi_\epsilon)$ is independent of ω_i and $\Psi_\epsilon = \mathrm{diag}(\tau_1^2, \ldots, \tau_p^2)$ is a $p \times p$ diagonal matrix (Zhang et al. 2016).

The non-recursive structural equation models and the identified SEM model can be shown as

$$y = By + \Gamma x + \zeta \leftrightarrow y = (I - B)^{-1}(\Gamma x + \varsigma)$$

The reciprocal relationships are described by the $p \times p$ matrix $B = (b_{ij})$ and the diagonal elements of B are fixed at 0. The $p \times q$ matrix Γ consists of the regression coefficients from x to y, $I - B$ is assumed to be non-singular, and the maximum of the absolute eigenvalue of B is less than 1. The $p \times 1$ vector ζ comprises random terms which are usually assumed to be independent of x, and we denote $\mathrm{Cov}\,[\zeta] = \Psi$, $\mathrm{Cov}\,[x] = \Phi$. The implied variance and covariance matrix derived can be expressed as

$$\sum(\theta) = \begin{matrix} (I - B)^{-1}\{\Gamma\Phi - B)\Gamma' + \Psi)(I - B)^{-1'} & (I - B)^{-1}\Gamma\Phi \\ \Phi\Gamma'(I - B)^{-1'} & \Phi \end{matrix}$$

where the parameter θ consists of all the elements of B, Γ, Φ, Ψ (Nagase and Kano 2017).

Evermann and Tate (2016) studied the SEM model and found that it could be applied for predictive analytics. In the paper, the author pointed out that the SEM model could have both manifest predictors and predicted manifest variables. Precisely, all exogenous latent variables were specified formatively, whereas all endogenous latent variables were specified reflectively. The model can be illustrated as in Fig. 1.

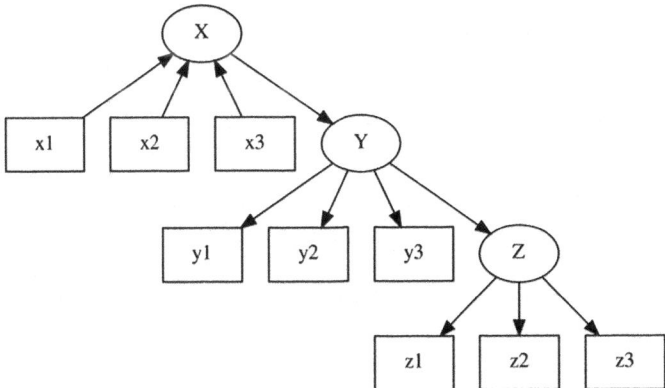

Fig. 1 Exogenous formative construct with manifest predictor variables

2.1.2 Pros and Cons of Structural Equation Modelling

Kaplan (2001) pointed out that there were some pros and cons regarding the SEM approach. The advantages include the potential for addressing critical substantive questions, the availability and simplicity of software dedicated to structural equation modelling, etc., while the cons include the attempt to attain a 'well-fitting' model, the post hoc justification for the ill-fitting model, etc. Fabrigar et al. (2010) indicated that the SEM approach had been applied in consumer psychology and related disciplines for over 30 years. Tsao et al. (2017) used the structural equation modelling approach to study the fatigue of Chinese railway employees and its influential factors. They used the SEM approach to analyse the contribution of each influential factor relating to fatigue.

3 Data Collection and Data Analysis

Construction accidents lead to different levels of injury to various parts of the body, the head, limbs, chest, back and neurologic injury. The level of injury is an unmeasurable matter, which translates to a latent variable in SEM. The levels of contributory negligence are estimated in the model and tested—to assess how it affects the compensation. This is about the breach of duty which can be attributed to the plaintiff at the time of the accident. Contributory negligence means that the plaintiff did not take due care. From the economic theory of tort, contributory negligence occurs when the marginal cost of taking care does not equal the decrease in the expected damage. If contributory negligence exists, a worker has to bear some responsibility for an accident and thus share the compensation with the defendant. The amount of compensation will be lower when there is contributory negligence. Therefore, it reflects the level of contributory negligence.

An initially specified base model with a logical path distribution between 14 variables was constructed (see Fig. 2). The variables with differences about their direction were assigned as a result of the general conceptualisation of injury in construction accident tort cases, and the tort compensation judgment classification. This model represents the predominantly compensation-oriented assessment process of tort construction accidents in Hong Kong by considering various factors that would be looked at during the judicial judgment procedure.

The model shows 14 variables. 13 of them, manifest variables; these latter are as follows: the gender of the plaintiffs (Gender), the age of the plaintiffs (Age), the monthly earnings of plaintiff before the accident (ME), the contributory negligence estimate (CL), the future loss of earnings (Future loss), the loss of earning capacity (LossofEC), the PLSA—pain, suffering and loss of amenities, the pretrial loss of earing (Pretrial), the special damage (SD), future treatment (FT), the death of plaintiff as a result of the accident (Death), the duration of judgment (Duration) and the level of the court (LC). The latent variable included in the model is the level of the injury that the plaintiff suffered from the construction accident (inj). The observed variables and

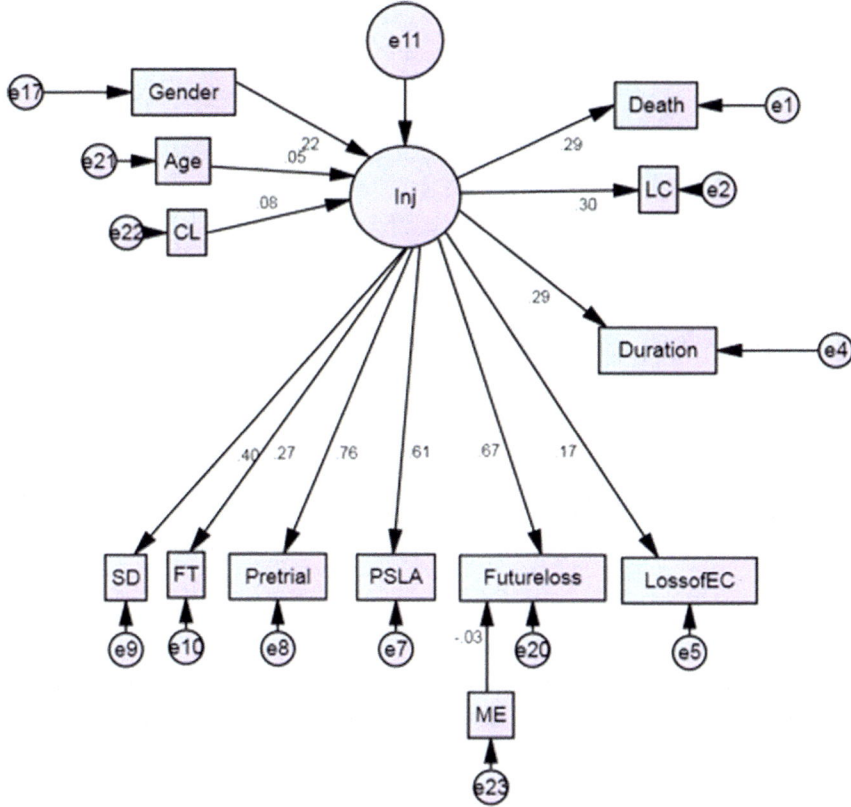

Fig. 2 Base model (1)

the latent variables were chosen to reflect the measurement of the effect of workplace accidents as related to the compensation paid.

First, Table 1 identifies the statistical descriptions of the exogenous variables and also provides the ranges of the variables with ordinal measurements. The ordinal variables are clearly stated along with their direction of coding. The duration of judgment is measured by year ranging from 3 years to 10 years. Gender is presented as an ordinal number, where 0 refers to a female plaintiff and one refers to a male. The ages of plaintiffs range from 18 years old to 65 years old. The level of court involved in these construction accident cases was assigned a range from 1 to 3, whereby 1 indicates the Court of Final Appeal (CFA); 2 represents the High Court (HC), as the Court of Appeal and the Court of First Instance; and finally, 3 indicates a District Court (DC).

The status of the cases is represented by 1 if the plaintiff won in court and 0 if the plaintiff lost in court. For the rest, the factors are all monetary variables with the unit being the Hong Kong dollar; this applies to employees' compensation confirmation, PSLA, future loss of earnings, loss of earning capacity, monthly earnings before

Table 2 Descriptive statistic for variables in base model

Variables	Mean		Range
Gender	N/A	0 (Female)	1 (Male)
Death	N/A	0 (Alive)	1 (Death)
Contributory negligence	N/A	1 (Yes)	0 (No)
Level of court	N/A	1 (CFA)	3 (DC)
Age	40.8	19	65
Monthly earning before accident	$14,273	$4400	$58,575
Future loss of earning	$496,619	$47,040	$3,990,060
Loss of earning capacity	$42,510.96	$0	$1,659,168
PSLA	$247,921.6	$0	$2,250,000
Pretrial loss of earning	$382,583.2	$5460	$2.335.904
Special damage	$26,723.7	$0	$1,036,439
Future treatment	$35,535.7	$0	$2,319,040
Duration of judgment	4.42 years	2 years	10 years

the accident, pretrial loss of earnings, special damage, future treatment and total compensation (Table 2).

The paths are initially assigned among the 14 variables, according to the rationale behind court cases in Hong Kong and by logical deduction. Compensation components are closely related to the injury level caused by accident. In the legal process, lawyers or judges examine the background of the accident through the oral evidence provided by both the worker and the employer. The medical report is a crucial factor in evaluating the level of injury, which affects future losses, medical expenses and funeral expenses in case of fatal accident compensation. 'Injury' is assigned a path to all compensation items. The physical status is influenced by the age, the gender of the worker and the effect of the injury level. Therefore, there is a path between injury, age and gender.

In the base model, the level of the future loss of earnings is assumed to have a relationship to the monthly earnings of the plaintiff before the accident. The worker's physical status is their key 'capital' in generating income from their work—which generally relies heavily on physical effort. If they have suffered physically from an injury, their productivity diminishes and this will, therefore, affect their wage rate. The court usually collects the information about the average monthly earnings of the workers before the accident to evaluate their working ability and to consider the effects of the injury on their productivity to provide a judgment on the compensation items: 'future loss of earning' and 'loss of earning capacity'. When the plaintiff has claimed compensation through ordinance initially, the amount of 'employees' compensation' depends on the monthly earnings of the plaintiff before the accident and this becomes a critical element that affects the compensation judgment in the court.

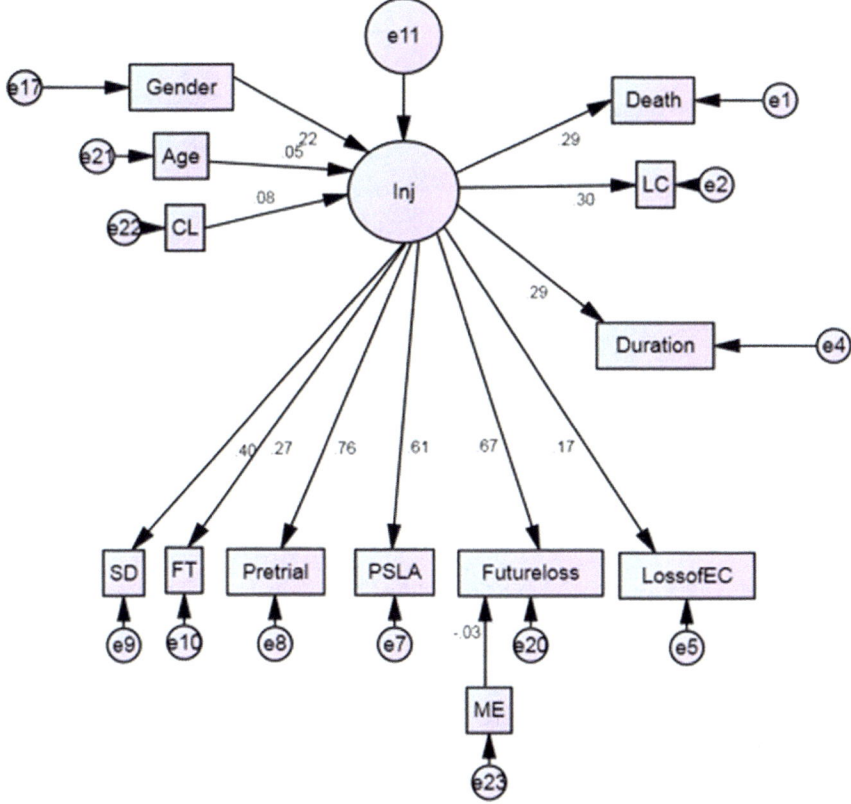

Fig. 3 Base model

The more complex the compensation judgment, the longer the time taken to produce the final judgment settlement. The severity of the injury will affect this duration. Equally, whether the plaintiff died or not in the accident, it will also be a result derived from the level of injury. Essentially, the factor of the contributory negligence of the plaintiff is a fundamental determinant in compensation judgments since externalities caused by the plaintiff's actions at the time of the accident should be considered (Fig. 3 and Table 3).

It can be concluded that the base model was significant, eight variables showed a statistically significant path such as the gender of the victims ($r^2 = 0.126, p < 0.05$), the level of injury on the duration of claim ($r^2 = 0.544, p < 0.01$) and the level of injury on death ($r^2 = 0.506, p < 0.01$). Regarding the level of compensation, most compensation categories such as PSLA, pretrial loss of earning and future treatment reflect on the level of compensation (Table 4).

The overall model fit of the base model estimates shows that RMSEA $= 0.099$ and CFI $= 0.603$. This indicates that the good fit of the base model is just at a moderate

Table 3 Parameter estimate of base model

Base model							
			Estimate	Estimate	S.E.	C.R.	P
Inj	<—	Gender	0.126	0.224	0.051	2.489	**0.013**
Inj	<—	Age	0.029	0.05	0.051	0.566	0.572
Inj	<—	CL	0.113	0.084	0.118	0.962	0.336
Death	<—	Inj	0.506	0.289	0.16	3.153	**0.002**
LC	<—	Inj	0.516	0.295	0.16	3.22	**0.001**
Duration	<—	Inj	0.544	0.293	0.17	3.196	**0.001**
SD	<—	Inj	0.307	0.396	0.073	4.23	***
FT	<—	Inj	0.166	0.272	0.056	2.979	**0.003**
Pretrial	<—	Inj	1.163	0.76	0.171	6.801	***
PSLA	<—	Inj	0.752	0.615	0.123	6.133	***
Futureloss	<—	Inj	1	0.669			
LossofEC	<—	Inj	0.115	0.172	0.06	1.913	0.056
Futureloss	<—	ME	−0.026	−0.025	0.066	−0.388	0.698

$***p < 0.0001$

level. Such a fit can help to indicate the correlation between a set of variables. A modified model is needed for more accurate interpretative purposes, as shown in Fig. 4.

Table 5 reveals the good fit of the modified model, which demonstrates a significant improvement. The value of CFI is now 0.946 (1 is the highest) which is even higher than the cut-off value and implies that this model is a perfect fit. Besides, although RMSEA also becomes lower, from 0.099 to 0.05, the model is still a close fit for explaining the construction accident compensation issue.

The parameter estimations are presented in Table 6. In the analysis of direct and indirect effects of those observed, which are exogenous variables, it was shown that death ($r^2 = 0.413, p < 0.05$) and duration ($r^2 = 0.457, p < 0.05$) were both significantly associated with the level of injuries, post-construction accident. On the other hand, variables that had a significant effect on the post-accident level of compensa-

Table 4 Good-fit index of the base model (1)

CMIN/DF	RMR	GFI	AGFI	TLI	CFI	RMSEA
2.577	0.074	0.863	0.808	0.523	0.603	0.099

Table 5 Good-fit index of the base model (2)

CMIN/DF	RMR	GFI	AGFI	TLI	CFI	RMSEA
1.405	0.027	0.95	0.912	0.921	0.946	0.05

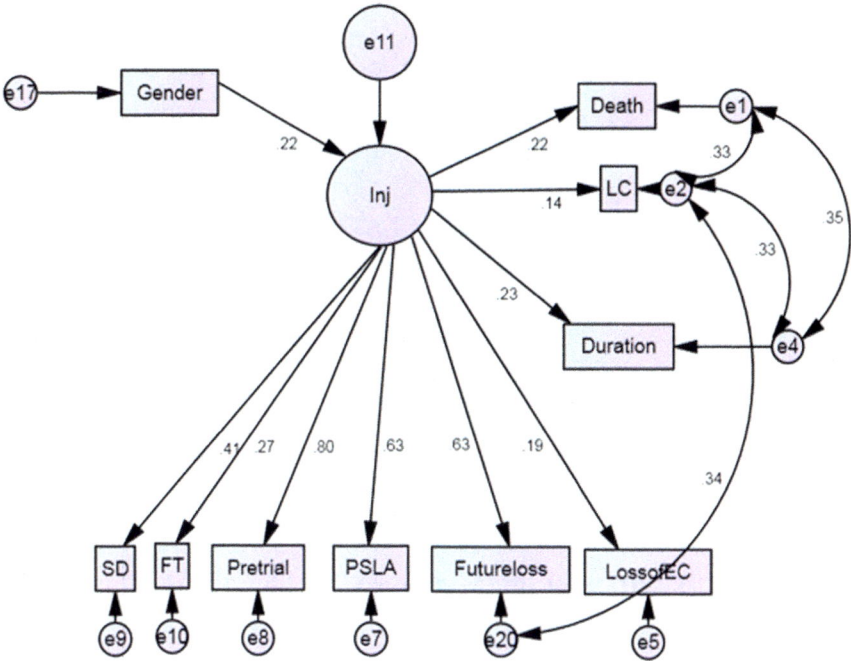

Fig. 4 Modified model

tion were those which were simply different components of the compensation: loss of EC ($r^2 = 0.136$, $p < 0.05$), future loss ($r^2 = 1$), PSLA ($r^2 = 0.822$, $p < 0.0001$), SD ($r^2 = 0.335$, $p < 0.0001$) and the highest pretrial loss of earnings ($r^2 = 1.293$), with p-value smaller than 0.0001.

The results found that the amount of the future loss of earnings was directly related to the level of injury. The more severe the injury level caused by accident, the more the compensation for the future loss. The employees suffered various physical injuries which affected their labour productivity. The average productivity of a worker affects their income level. The injury affects the human capital related to labour and future income. Higher levels of injury lead to a decline in productivity, and higher compensation should be used to 'compensate' the injured employee. The level of injury exhibits a significant positive effect on the duration of judgment. When a worker suffered from complicated or multiple injuries, which needed more time to heal, this increased the time necessary for obtaining a medical report, and for evaluating the compensation that should be paid to the plaintiff to cover the losses.

Table 6 Parameter estimate of the modified model

Modified model							
			Estimate	Estimate	S.E.	C.R.	P
Inj	<—	Gender	0.115	0.216	0.048	2.392	0.017
Death	<—	Inj	0.413	0.222	0.169	2.44	0.015
LC	<—	Inj	0.252	0.136	0.16	1.575	0.115
Duration	<—	Inj	0.457	0.231	0.18	2.538	0.011
SD	<—	Inj	0.335	0.406	0.078	4.274	***
FT	<—	Inj	0.177	0.272	0.059	2.975	0.003
Pretrial	<—	Inj	1.293	0.795	0.198	6.545	***
PSLA	<—	Inj	0.822	0.632	0.135	6.086	***
Futureloss	<—	Inj	1	0.632			
LossofEC	<—	Inj	0.136	0.192	0.064	2.124	0.034

*** $p < 0.0001$

4 The Contribution of the Model

Research that has studied the relationship between injury and compensation by using court cases is scarce. A few studies examine only the economic meaning and implications of tort but not benefit. This book chapter highlights the various factors relating to accident compensation claims. These empirical findings may enhance policymakers' awareness regarding the compensation that takes no account of the plaintiffs' responsibilities for, and their non-financial losses regarding the accident. The plaintiffs must seek another scheme for compensating their damage, which will in turn increase the legal fees which must be paid. The government could consider the need to provide compensation, in relation to both financial and non-financial damages, to the injured employee to alleviate the pressure on the legal system in Hong Kong.

5 Limitation of Constructing SEM

The first limitation of the structural equation modelling analysis in this study was the theoretical issues between the variables. There have been no previous economic studies in the literature which have discussed the relationship between the compensation items concerning accidental injuries. Some research has addressed the connection between the compensation system and the injury problem only but does not divide the compensation provided into various categories for examination purposes. Alternatively, in legal research, law experts have mainly explained the rationale behind the judgment qualitatively. None of the legal research studies appears to have looked at compensation, on a case-by-case basis, quantitatively. Thus, this presents an obstacle concerning constructing the relationships between variables to build a base model. Second, the cross-sectional nature of the measured variables is also a limitation. The court cases collected have been of a similar view since all cases were related to construction accidents in Hong Kong.

Acknowledgements An earlier version of the paper was presented in European Real Estate Society Conference 2017.

References

Angelini, M.E., and G.B.M. Heuvelink. 2018. Including spatial correlation in structural equation modelling of soil properties. *Spatial Statistics* 25: 35–51.

Azhar, S. 2017. Role of visualization technologies in safety planning and management at construction jobsites. *Procedia Engineering* 171: 215–226.

Bunting, J., C. Branche, C. Trahan, and L. Goldenhar. 2017. A national safety stand-down to reduce construction worker falls. *Journal of Safety Research* 60: 103–111.

Chen, Y., B. McCabe, and D. Hyatt. 2017. Impact of individual resilience and safety climate on safety performance and psychological stress of construction workers: A case study of the Ontario construction industry. *Journal of Safety Research* 61: 167–176.

Chen, Y., B. McCabe, and D. Hyatt. 2018. A resilience safety climate model predicting construction safety performance. *Safety Science* 109: 434–445.

Evermann, J., and M. Tate. 2016. Assessing the predictive performance of structural equation model estimators. *Journal of Business Research* 69 (10): 4565–4582.

Fabrigar, L.R., R.D. Porter, and M.E. Norris. 2010. Some things you should know about structural equation modeling but never thought to ask. *Journal of Consumer Psychology* 20 (2): 221–225.

Glofcheski, R. 2012. *Tort Law in Hong Kong*, 3rd ed. Hong Kong: Sweet & Maxwell.

Guo, H., Y. Yu, and M. Skitmore. 2017. Visualization technology-based construction safety management: A review. *Automation in Construction* 73: 135–144.

Jöreskog, K.G. 1973. A general method for estimating a linear structural equation system. In *Structural equation models in the social sciences*, ed. A.S. Goldberger, O.D. Duncan, 85–112. Academic Press.

Kaplan, D. 2001. Structural Equation Modeling A2—Smelser, Neil J. In *International encyclopedia of the social & behavioral sciences*, ed. P.B. Baltes, 15215–15222. Oxford: Pergamon.

Li, R.Y.M. 2015. Construction safety knowledge sharing via smart phone apps and technologies. In *Handbook of mobile teaching and learning*, 261–273.

Li, R.Y.M., and S.W. Poon. 2009. Workers' compensation for non-fatal accidents: Review of Hong Kong Court cases. *Asian Social Science* 5 (11): 15–24.

Li, R.Y.M., and S.W. Poon. 2011. Using Web 2.0 to share knowledge of construction safety: the fable of economic animals. *Economic Affairs* 31 (1): 73–79.

Li, R. Y. M., W. Chau Kwong, and W. Ho Daniel Chi. 2017a. Dynamic panel analysis of construction accidents in Hong Kong. *Asian Journal of Law and Economics*.

Li, Q., C. Ji, J. Yuan, and R. Han. 2017b. Developing dimensions and key indicators for the safety climate within China's construction teams: A questionnaire survey on construction sites in Nanjing. *Safety Science* 93: 266–276.

Liao, C.-W., and T.-L. Chiang. 2015. The examination of workers' compensation for occupational fatalities in the construction industry. *Safety Science* 72: 363–370.

Liu, J., and C.-Z. Hu. 2017. Application of information technology in active safety control for construction machine. *Procedia Engineering* 174: 1182–1189.

Misiurek, K., and B. Misiurek. 2017. Methodology of improving occupational safety in the construction industry on the basis of the TWI program. *Safety Science* 92: 225–231.

Mohammad, M.Z., and B.H.W. Hadikusumo. 2017. A Model of integrated multilevel safety inter-vention practices in Malaysian construction industry. *Procedia Engineering* 171: 396–404.

Nagase, M., and Y. Kano. 2017. Identifiability of nonrecursive structural equation models. *Statistics & Probability Letters* 122: 109–117.

Park, R.M., and A. Bhattacharya. 2013. Uncompensated consequences of workplace injuries and illness: Long-term disability and early termination. *Journal of Safety Research* 44: 119–124.

Schreiber, J.B. 2008. Core reporting practices in structural equation modeling. *Research in Social and Administrative Pharmacy* 4 (2): 83–97.

Torres, L.D., and A. Jain. 2017. Employer's civil liability for work-related accidents: A comparison of non-economic loss in Chile and England. *Safety Science* 94: 197–207.

Tsao, L., J. Chang, and L. Ma. 2017. Fatigue of Chinese railway employees and its influential factors: Structural equation modelling. *Applied Ergonomics* 62: 131–141.

Wang, J., and X. Wang. 2012. *Latent growth models for longitudinal data analysis.*

Zhang, Y.-Q., G.-L. Tian, and N.-S. Tang. 2016. Latent variable selection in structural equation models. *Journal of Multivariate Analysis* 152: 190–205.

Chapter 8
Green Roof Safety: A Tale of Three Cities

Abstract Green roofs have immense benefits in large cities. They substantially lower the temperature of urban cities which is increased due to high-rise buildings; they reduce the problem of the heat island effect. Therefore, many urban planners and environmentalists advocate the construction of green roofs on the tops of buildings. The green roof collapses in Hong Kong, Illinois and Latvia remind us of the importance of proper planning and management in green roof construction. Green roofs are usually associated with heavy soil layer, and rainfalls and snowfalls increase the weight of the substances on the green roof; this is what has led to their collapse. In this chapter, we review the structure, safety and sustainability benefits of green roofs. Three case studies from Hong Kong, Illinois and Latvia are included for illustration purpose.

Keywords Green roof · Sustainability · Case studies

1 Introduction to Green Roofs

There are three major types of green roof: intensive, semi-intensive and extensive green roofs. Intensive green roofs are characterised by thick substrate layers which require significant maintenance; the extensive green roofs have shallow substrate layers and require less maintenance; semi-intensive green roofs have moderately thick substrate layers and can accommodate small herbaceous plants, grasses, etc. (Vijayaraghavan 2016).

Many green roofs have similar structures to those shown in Fig. 1. The upper layer is of vegetation (including plants, flowers, etc.), planted by the occupant. Under this layer of vegetation, there is the growing medium layer (the depth of the ever-increasing medium changes according to the design specifications). A layer of water retention fabric is optionally placed under the layer of growing medium (e.g. the soil layer). The bottom layer in Fig. 1 is the drainage layer. It is placed over a root barrier that covers the roofing membrane. Between the drainage layer and the layer of water retention fabric, there is a layer of filter fabric, which keeps silt and particulate matter in the growing medium from clogging the drainage layer (Getter and Rowe 2006).

© Springer Nature Singapore Pte Ltd. 2019
R. Y. M. Li, *Construction Safety Informatics*,
https://doi.org/10.1007/978-981-13-5761-9_8

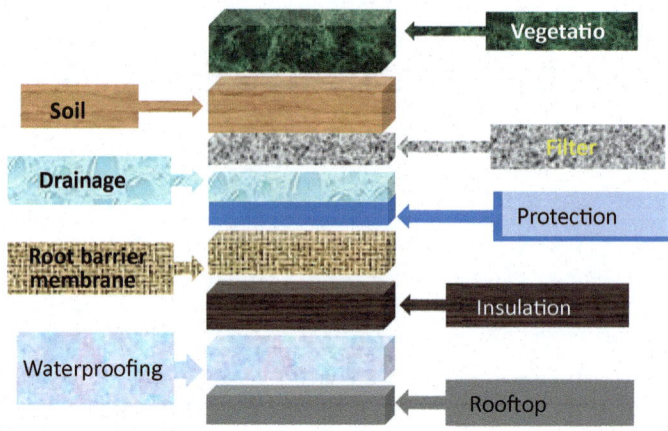

Fig. 1 Typical components of a green roof (Shafique et al. 2018)

Green roofs have always been considered as a sustainable building structure which can achieve the three major desiderata of sustainability, i.e. the environmental, social and economic requirements (Li and Pak 2010; Li 2013; Li and Li 2018). Roof greening alleviates the heat island effect brought about by extensive areas of walled buildings in urban areas (Li et al. 2018). Besides, while many cities' governments aim to reduce the greenhouse gas emission, for example, the Chinese government committed to reducing carbon dioxide, green roofs can be a good solution in this regard (Liu et al. 2015).

Extensive and intensive green roofs (Architectural Services Department 2007)

	Extensive green roof	Intensive green roof
Characteristics		
Soil	• Thin (50–150 mm thick)	• Deep (200–2000 mm thick)
Irrigation	• Little or no irrigation	• Irrigation is needed
Maintenance fee	• Low	• Higher
Area of application	• Extensive	• Intensive
Advantages		
Weight	• Lightweight	• Diverse utilisation of roof for recreation, growing food and open space
	• Low maintenance cost	• Greater plants and habitats diversity
	• Suitable for retrofit projects and large areas	• Good insulation

(continued)

(continued)

	Extensive green roof	Intensive green roof
	• Suitable for roofs with 0–30 degree of slope	• Can simulate a wildlife garden can be very attractive
	• Allow vegetation to develop naturally	• Visually accessible
Disadvantages	• Limited choice of plants	• Relatively higher cost
		• Not suitable for green roof retrofit projects
	• Usually no access for recreation or other uses	• Greater weight loading
	• Visually unattractive in a dry season	• Need for irrigation and drainage systems, and hence greater need for energy, water materials, etc

2 History of Green Roof

Having a green roof is an ancient technique. Latter, humans constructed the green roofs because rooftop gardens have good insulation characteristics and reduce the adverse effects of urbanisation. One of the most well-known ancient green roofs was, in fact, the Hanging Gardens of Babylon constructed around 500 BCE. Modern green roofs started in Germany in the early 1960s, and the electricity crises motivated their construction. Germans began building green roofs to reduce the energy consumption of their homes. Germany is the leader in modern green roof construction, because green roofs have been built in Germany, on a huge scale, and have been advanced regarding design, etc. since those early years in the 1960s. In 1962, a German researcher, Reinhard Bornkamm, published his book on green roof research. New developments in roof gardens initiated with the aid of BDLA at the Deubau exchange were used in Essen in 1973. In the early 80s, the green roofs market expanded quite quickly, and many roofs were built in Germany by inexperienced developers. Forschungsgesellschaft Landschaftsentwicklung Landschaftsbau published green roof building guidelines in German. These guidelines were released with the aid of the association of standards and testing materials in 2005 and 2006, wherein the construction of green roofs was described in detail, including the planning, execution and protection of green roofs (Shafique et al. 2018).

In 2009, USEPA released information on the construction and advantages of green roofs. Research into the green roofs guidelines and the renovation and management of green roofs were carried out in the US and shared with other countries. At present, cities in countries like the US, Canada, Singapore, Australia, Japan, China, Hong Kong and South Korea are pursuing robust initiatives to use the green roofs to achieve the multiple advantages that they can offer (Shafique et al. 2018).

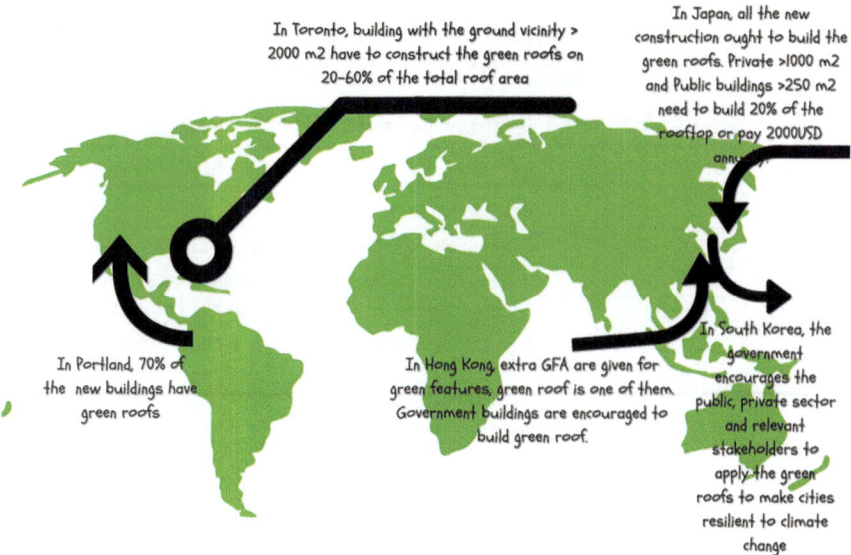

In Toronto, building with the ground vicinity > 2000 m2 have to construct the green roofs on 20–60% of the total roof area

In Japan, all the new construction ought to build the green roofs. Private >1000 m2 and Public buildings >250 m2 need to build 20% of the rooftop or pay 2000USD annually.

In Portland, 70% of the new buildings have green roofs

In Hong Kong, extra GFA are given for green features, green roof is one of them. Government buildings are encouraged to build green roof.

In South Korea, the government encourages the public, private sector and relevant stakeholders to apply the green roofs to make cities resilient to climate change

Fig. 2 Important milestone of green roof development worldwide from 1960 to 2018 (Shafique et al. 2018)

In Toronto, buildings with a ground area of more than 2000 m^2 are required to have green roofs on 20–60% of the total roof area. In Japan, it is recommended that all new construction should have green roofs. Private building > 1000 m^2 and public buildings > 250 m^2 need to have 20% of the rooftop as a green roof or pay 2000USD annually. In Portland, in the United States of America, 70% of the roof area of any new building must consist of a green roof. In Portland, there were approximately 2 acres of a green roof in 2005, and additionally, a larger area of green roof was planned to reap the sustainable benefits that such roofs offer. In China and Hong Kong, the governments recommended that green roofs should be used. In South Korea, the authorities are encouraging the public, individual developers and stakeholders to use green roofs to make cities safe, sustainable and resilient to a variety of weather conditions (Shafique et al. 2018) (Fig. 2).

3   Benefits of Green Roofs

3.1   Financial Benefits

As compared with the traditional concrete rooftop, there are some financial benefits of using green roofs. Niu et al. (2010) suggested that green roofs have lower maintenance costs, save energy and they absorb carbon dioxide. Carter and Keeler (2008) found that the net present value of using green roofs was higher than that of using normal roofs in the Tanyard Branch Watershed, the US, which meant that a green roof's financial benefits exceeded those of a traditional roof.

3.2 Social and Environmental Benefits

There are some advantages of using green roofs. Dimitrijevic et al. (2016) suggested that a green roof not only provides thermal comfort for the residents but also reduce the energy consumption of buildings. Berardi (2016) indicated that, in his study, he found that green roofs could significantly reduce the temperature of urban areas and this might be a strategy to help to mitigate the heat island effect. Yang et al. (2008) undertook a research study in Chicago and found that green roofs could significantly reduce air pollution.

By studying the green roofs installed in some high-density urban areas, in Canada, the US, the UK, Thailand and Taiwan, (Hui 2011) concluded that green roofs could play an important role in sustainability, especially when green roofs are used for urban farming. Although he argues that Hong Kong has huge potential for the increase in green roofs, due to their environmental benefits, he also contends that technical issues, local conditions, guidelines and practical experiences are essential and necessary things to take into account when we engage in rooftop urban farming.

Similarly, Yan (2011) summarised the benefits of green roofs as follows: they help regulate climate, improve environmental quality, reduce the city heat island effect, fix carbon, release oxygen, purify the air and conserve energy. (Ma et al. 2012) experimented Jinan and found that green roofs help rainwater retention and solve the problem of rainwater run-off in urban areas.

3.3 Other Benefits

In addition to the financial and social benefits, green roofs have other advantages. Szlivka and Zachar (2015) suggested that green roofs could make buildings have an extra aesthetic appeal, and thus provide people with a pleasant experience. Hui (2006) indicated that in Hong Kong, the use of green roofs could bring a great many benefits to the city and these benefits included an increase in the roofs' life expectancy, better use of space, a reduction in the risk of glare about surrounding buildings, etc.

4 Disadvantages of Green Roof

The green roof also has some disadvantages. First, the initial cost of a green roof is much higher than that of a traditional roof. In the US, even the cheapest green roof installation was \$108–248/m^2 (\$10–23/ft^2), double the price of other roof types (Sproul et al. 2014). Zhang et al. (2012) found that there were some barriers to the adoption of green roofs in Hong Kong, such as a lack of promotion from the government and an increase in maintenance costs.

Clayden and Osmundson (2001) point out that if green roofs are designed or maintained poorly, they may collapse due to overloading. Moreover, (Magill et al. 2011) pointed out that weight is a crucial issue when constructing green roofs. If such a roof is not provided with sufficient support and there is inadequate planning of the green roof system, it is likely that a serious collapse will occur. Besides, Vijayaraghavan (2016) stresses the disadvantages of garden soil on green roofs; the use of garden soil is the major factor in the risk of collapse.

By compiling a database of 526 commercial buildings in the Melbourne CBD area, and using bivariate analysis, Wilkinson and Reed (2009) found that due to the various requirements for green roofs, such as local climatic conditions, rainfall patterns and the construction of the physical property stock, approximately only 15% of existing buildings which are suitable for green roofs construction.

5   Case Studies

A case study is the best tool for social analysis when (1) we attempt to ask questions of why, (2) when the research question emphases on a real-life context and (3) when the research questions require an in-depth study. These three-criteria fit our research well. Besides, case studies enable the researchers to attain in-depth results by assuming multiple sources of data…new insights can be found in this process that eventually adds breadth, rigour and depth (Li 2014).

5.1   City University of Hong Kong: The Collapse of the Green Roof

Yuen (2016a) reported that on 20 May 2016, a 35 × 40 m roof collapsed at around 2:30 pm—at the Hu Fa Kuang Sports Centre building. The accident on the 20th May injured three people (Cheng and Cheung 2016) (Fig. 3).

To investigate the incident, a nine-person Investigation Committee (IC) was appointed by the President of the City University of Hong Kong on 23 May 2016 (CityU 2016). The IC of CityU concluded that the design of the green roof, especially the loading assessment (coupled with the apparent use of incorrect data/information), was likely the primary, core factor which caused the accident (CityU 2016).

After the collapse of the green roof in CityU, the other universities in Hong Kong started to check their roofs to make sure such an accident could not happen on their premises. The Chinese University of Hong Kong (CUHK) issued a guideline for

Fig. 3 The collapse of green roof at Hong Kong City University (Yau et al. 2016)

green roof safety for their students and announced that they would remove vegetation grown on six outbuildings (including laboratories, the gym room, the school library, etc.) next to its sports field (Tsang 2016).

5.1.1 Public Opinions About the Collapse of the Green Roof at Hong Kong City University

There are some views in the local community about the events in Hong Kong CityU. After the incident, Yang Ming, a Ph.D. candidate in CityU said that he hoped the president would go on TV to say something to the students, to present a responsive image. Besides, an instructor at CityU advised that the university should make plans to raise safety awareness among students and teachers and she also believed that if there were more significant number of students, this would make escape more difficult (Li 2016).

5.1.2 Professional Perspectives on the Collapse of the Green Roof at Hong Kong City University

In the following, we present some professional views on the events in Hong Kong CityU. Yim Kin-ping, Chairman of the Hong Kong Project Management Exchange Centre said that the collapsed sports centre was more than 20 years old. There was a need to make thorough plans when green roof construction is envisaged and a rigorous examination of the load-bearing capacity of the building, possibly leading to its strengthening should be carried out. Otherwise, problems and tragedies will occur, especially when there is heavy rainfall (Li 2016). The Secretary for Education Eddie Ng Hak-kim said that the collapse of the green roof in CityU seemed to be an isolated incident, but it did affect the other eight universities (Yuen 2016b). The education sector lawmaker Ip Kin-yuen said that he found the report was acceptable but needed to undertake separate investigations with the building department to clarify what types of disciplinary actions should be taken regarding the responsible parties. Besides, John Tse Wing-ling, chairman of the CityU Staff Association and also a council member, understood that the human resources office should also be blamed for their failure to ensure that the office had enough professionals to supervise such a project (Yeung 2016).

5.2 Green Roof Collapses in Illinois

According to the news, a $500' \times 60'$ section of green roof collapsed in St. Charles, Illinois, and this caused $13 million of damage. But, fortunately, it did not lead to any deaths. The consequent court suit stated that 'the structure was not appropriately designed and constructed' (Hitzeman 2013). The city had snowfall, and then the temperatures climbed above freezing. The collapse happened at the beginning of the thaw; there had been an ice accumulation on the roof, and this had ultimately caused the rooftop collapse. This might have been prevented if the meltwater had been able to drain off in a better way (Fountain 2011).

5.3 Green Roof Collapses in Latvia

Appl (2014) reports that a green garden was being constructed on a supermarket roof and that this led to the collapse of the supermarket roof. This happened in Riga, the capital of Latvia, and 54 people died in this accident. Many factors conspired to make the incident more dangerous regarding loss of lives. Customers were discouraged from leaving the store with unpaid goods by the security guards at the checkout counters after an alarm sounded. The customers queued up to pay for their goods underneath the roof which was collapsing. The supermarket checkout staff were instructed to ignore the alarm (Woolfson and Vanadzins 2014). They believed that the alarm had been falsely triggered by welding construction work in the basement as some false alarms had plagued the store prior the roof collapse. In the absence of evidence of a fire, the alarm was switched off by a technician who was called out to inspect it (Woolfson and Juska 2014a). Staff had been told to remain at their posts to serve the customers and deal with their purchases (Woolfson and Vanadzins 2014).

The service counter staff were instructed to protect the merchandise under their supervision even after the collapse. Fleeing customers and staff tried to leave by side doors but then found that these had locked automatically after the loss of electrical power. Trapped survivors began to break the store's glass windows to leave the shattered building. A lack of emergency provisions and inadequate emergency instructions, coupled with the blocking of emergency exits due to stacked supermarket products added to the death toll (Woolfson and Vanadzins 2014).

Technical investigations and speculations about the causes of the store collapse focused on faulty design, builders' negligence or some combination of the two. The three Latvian architects who reviewed the building design found that the collapse happened due to it being designed with insufficient load-carrying capacity—three times less than what was required. Riga Technology University experts concurred but focused more on the roof truss connection screws which were chosen; these were of lesser carrying capacity than was needed, and of inferior quality (Woolfson and Juska 2014a). Moreover, Woolfson and Juska (2014b) argued that it is necessary to establish accountability for social harms to make it less likely that a similar accident could occur again.

6   Conclusion

Green roof is beneficial to our modern walled cities as that provides the vegetation in city centre and aesthetical impact to us. Nevertheless, that poses difficulty to facility management staff as they have to be aware of the accumulation of water which leads to green roof collapse. Structural safety issues, highlighted by the previous incidents in Illinois, Latvia and Hong Kong, have to be considered seriously by these staff. While that may be a threat to the existing green roof owners and contractors, it can also be an opportunity for the facility managers and building surveyors to exercise their expertise in providing professional and timely advice to our stakeholders. Risk management after and during the rooftop collapse is equally important. Training of facility management staff is therefore important in this regard.

References

Appl, R. 2014. *Riga supermarket collapse: Who's to blame?* Available from http://www.igra-world. com/images/news_and_events/IGRA_Green_Roof_News_1_2014.pdf.

Architectural Services Department. 2007. *Study on green roof application in Hong Kong* 2007, 11 October 2018. Available from https://www.archsd.gov.hk/media/11687/1353-green-roofs-es-2007-02-16.pdf.

Berardi, U. 2016. The outdoor microclimate benefits and energy saving resulting from green roofs retrofits. *Energy and Buildings* 121:217–229.

Carter, T., and A. Keeler. 2008. Life-cycle cost-benefit analysis of extensive vegetated roof systems. *Journal of Environmental Management* 87(3).

Cheng, K., and K. Cheung. 2016. Roof collapses at City University sports centre, three injured. In *Hong Kong Free Express (HKFE)*.

CityU. 2016. University Response to the Report of the Investigation Committee for the CityU Sports Hall Incident. Retrieved from: http://www.cityu.edu.hk/ic/ua_0610.htm. [assessed on 6 March 2019].

Clayden, A., and T. Osmundson. 2001. Roof gardens: History design and construction. *Garden History* 29 (2):226–227.

Dimitrijevic, D., M. Tomic, P. Zivkovic, M. Stojiljkovic, and M. Dobrnjac. 2016. Thermal characteristics and potential for retrofit by using green vegetated roofs. *Annals Of The Faculty Of Engineering Hunedoara—International Journal Of Engineering* 14 (1):41–44.

Fountain, H. 2011. Green Roof Collapses in Illinois. *The New York Times.*

Getter, K.L., and D.B. Rowe. 2006. The role of extensive green roofs in sustainable development. *HortScience* 41 (5):1276–1285.

Hitzeman, H. 2013. *St. Charles firm sues for $13 M after green roof collapse in 2011.* Available from http://www.dailyherald.com/article/20130306/news/703069735/.

Hui, S. 2011. Green roof urban farming for buildings in high-density urban cities. In *The 2011 Hainan China World Green Roof Conference.* Hainan (Haikuo, Boao and Sanya).

Hui, S.C.M. 2006. Benefits and potential applications of green roof systems in Hong Kong. In *Proceedings of the 2nd Megacities International Conference 2006* 351–360.

Li, R.Y.M. 2013. The usage of Automation System in Smart Home to provide a Sustainable Indoor Environment: A Content Analysis in Web 1.0. *International Journal of Smart Home* 7(4):47–60.

Li, R.Y.M. 2014. Transaction costs, firms' growth and oligopoly: Case studies in Hong Kong real estate agencies' branch locations *Asian. Social Science* 10 (6):41–52.

Li, Y. 2016. CityU committee begins rooftop collapse probe. China Daily Asia.

Li, R.Y.M., and D.H.A. Pak. 2010. Resistance and motivation to share sustainable development knowledge by Web 2.0. *Journal of Information & Knowledge Management* 09(03):251–262.

Li, R., and H. Li. 2018. Have housing prices gone with the smelly wind? Big data analysis on landfill in Hong Kong. *Sustainability* 10 (2):341.

Li, R., K. Cheng, and M. Shoaib. 2018. Walled buildings, sustainability, and housing prices: An artificial neural network approach. *Sustainability* 10 (4):1298.

Liu, X., G. Mao, J. Ren, R.Y.M. Li, J. Guo, and L. Zhang. 2015. How might China achieve its 2020 emissions target? A scenario analysis of energy consumption and CO_2 emissions using the system dynamics model. *Journal of Cleaner Production* 103:401–410.

Ma, L., B. Qin, and C. Qingzuo. 2012. Performance of urban rainwater retention by green roof: A Case Study of Jinan. *Applied Mechanics and Materials* 178–181:295–299.

Magill, J.D., K. Midden, J. Groninger, and M. Therrell. 2011. *A history and definition of green roof technology with recommendations for future research.* cited. Available from http://opensiuc.lib. siu.edu/cgi/viewcontent.cgi?article=1132&context=gs_rp.

Niu, H., C. Clark, J. Zhou, and P. Adriaens. 2010. Scaling of economic benefits from green roof implementation in Washington. *DC. Environmental Science & Technology* 44 (1):4302–4308.

Shafique, M., R. Kim, and M. Rafiq. 2018. Green roof benefits, opportunities and challenges—a review. *Renewable and Sustainable Energy Reviews* 90:757–773.

Sproul, J., M.P. Wan, B.H. Mandel, and A.H. Rosenfeld. 2014. Economic comparison of white, green, and black flat roofs in the United States. *Energy and Buildings* 71:20–27.

Szlivka, D., and A. Zachar. 2015. Thermal measurement and calculation of green roof and normal flat roof. *Annals Of The Faculty Of Engineering Hunedoara—International Journal Of Engineering* 13 (3):109–112.

Tsang, E. 2016. Chinese University of Hong Kong to remove green roofs. Buildings Department Issues New Guidelines to Schools for Immediate Safety Checks. In *South China Morning Post.*

Vijayaraghavan, K. 2016. Green roofs: A critical review on the role of components, benefits, limitations and trends. *Renewable and Sustainable Energy Reviews* 57:740–752.

Wilkinson, S.J., and R. Reed. 2009. Green roof retrofit potential in the central business district. *Property Management* 27 (5):284–301.

Woolfson, C., and A. Juska. 2014a. Neoliberal Austerity and Corporate Crime: The Collapse of the Maxima Supermarket in Riga, Latvia. *New Solutions: A Journal of Environmental and Occupational Health Policy* 24 (2):129–152.

Woolfson, C., and A. Juska. 2014b. *Safety Crime in Neoliberal Post-communist Society: the Collapse of Maxima Supermarket in Riga, Latvia*. cited. Available from http://liu.diva-portal.org/smash/get/diva2:690949/FULLTEXT01.pdf.

Woolfson, C., and I. Vanadzins. 2014. Historical and contemporary challenges to occupational safety and health in Latvia. *Policy and Practice in Health and Safety* 12 (2):47–65.

Yan, B. 2011. The research of ecological and economic benefits for green roof. *Applied Mechanics and Materials* 71–78:2763–2766.

Yang, J., Q. Yu, and P. Gong. 2008. Quantifying air pollution removal by green roofs in Chicago. *Atmospheric Environment* 42 (31):7266–7273.

Yau, C., C. Lo, J. Lau, and S. Lau. 2016. Answers Demanded Over Collapse of Green Roof at Hong Kong City University. In *South China Morning Post*.

Yeung, R., ed. 2016. *Surveyor expected to be held liable for roof collapse as City University's top brass are accused of offloading responsibility*.

Yuen, C. 2016a. Construction Industry Figures Fail to Attend CityU Roof Collapse Investigation. In *Hong Kong Free Express (HKFE)*.

Yuen, C. 2016b. 'We are very concerned about structural safety,' says CY in first comment after CityU roof collapse.

Zhang, X., L. Shen, V.W.Y. Tam, and W.W.Y. Lee. 2012. Barriers to implement extensive green roof systems: A Hong Kong study. *Renewable and Sustainable Energy Reviews* 16 (1):314–319.

Chapter 9
Automated and Intelligent Tools
in the Construction Industry

Abstract Hong Kong is a city which has many high-rise buildings, civil engineering and refurbishment projects. In this chapter, we study the accident-prone factors that lead to accidents on building sites by interviewing various stakeholders. In the second part of the study, we extract the causes of accidents from court case reports. Finally, we include a single case study based on the activities of one of the largest construction companies in Hong Kong—who use AI chatbot for filling the accident report, smart concrete sensors, robotics, drone photogrammetry, three-dimensional (3D) printing, virtual reality, wearable robotics, etc. on their construction sites. We look at whether innovative tools can be useful in improving safety performance on construction sites.

Keywords Construction safety · AI chatbot · Smart concrete sensor · Holographic mixed reality · Drone photogrammetry · Infographics

1 Introduction

The lethal working environments which can often be found in the building industry have led to there being approximately 60,000 fatalities worldwide in this industry each year. This is higher than the number of deaths suffered by any other industry. Besides, mishaps affect not only workers' well-being, but also the quality of life of whole families (Alarcón et al. 2016). Some victims are breadwinners, and so accidents which harm them severely affect their families in several ways. Even though everyone involved wants construction safety performance to be improved, construction sites remain one of the most perilous workplaces overall, and a danger-free place is an exception rather than the rule (Geoff 2016). In recent decades, there has been a trend towards increasingly higher buildings, and this has increased the complexity of construction projects and the level of risk involved (Guo et al. 2015). Timely risk assessments are an absolute necessity concerning many construction projects (Guo et al. 2015).

Accidents on sites are hardly ever the result of pure bad luck. Accidents are not now viewed as arbitrary chance occurrences but instead as the consequence of a progression of events that could conceivably be identified and controlled (Alarcón

© Springer Nature Singapore Pte Ltd. 2019
R. Y. M. Li, *Construction Safety Informatics*,
https://doi.org/10.1007/978-981-13-5761-9_9

et al. 2016). A large proportion of these 'misfortunes' occur mostly because of the absence of visible safety measures: for example, safety nets and guide rails (Edrei and Isaac 2016, forthcoming; Li and Poon 2013; Li 2015). Temporary workers are regularly used to provide workforce when the time is short. These workers are at an increased risk of injury because many of them change job assignments several times a year, and they are not always trained or re-trained. Host employers are less willing to devote training resources to temporary workers. It is not uncommon for a temporary worker to suffer a fatality after just 1 or 2 days on the job (Mroszczyk 2015).

Previous research has shown that designers can play an essential role in ensuring construction safety (Bong et al. 2015). Indeed, accidents often happen because of oversights by designers and/or safety officers, which occur when they arrange or execute safety activities. The application of innovative initiatives can enhance the development of safety on sites (Edrei and Isaac 2016, forthcoming), and any commitment to reduce work-related accidents can be viewed as worthwhile (Alarcón et al. 2016).

Accidents originate from several organisational levels. However, a study of 1180 firms using 221 different working practices over 4 years in Chile demonstrated that working methods are the most critical factors that affect accident rates (Alarcón et al. 2016).

In this present study, we first review the factors which lead to accidents on construction sites from practitioners' and from trial judges' perspectives—via extensive qualitative interviews. We compare the factors mentioned by the groups of interviewees about high-rise and low-rise buildings. While the practitioners have a day-to-day work experience on construction sites, and their views can be subjective. On the other hand, the trial judges must consider all the evidence provided by the victims, employers, witnesses, literature, etc. and assess the factors which have led to the accidents they have to deal with. Hence, they can be considered more objective. Thus, the second aim will be to compare the practitioners and the judges' perceptions of the factors which commonly lead to accidents. Finally, in the era of information technology (IT), advances in various kinds of IT tool present an unprecedented opportunity to improve safety performance on construction sites (Skibniewski 2014; Hoła et al. 2015). We will use one single case study to demonstrate the application of innovative technological tools on construction sites. This should provide some idea of whether devices such as 3D printing, robots, aerial photogrammetry, etc. can reduce the chances of accidents on construction sites.

2 Safety Performance Indicators

Historically, the construction industry examines safety performance via lagging indicators; these include numbers of injuries, illnesses and fatalities. One noteworthy limitation of safety performance appraisal through lagging indicators is that incidents must happen before risks or hazardous conduct can be recognised. Leading indicators are safety measurements that proactively survey execution safety performance by gauging procedures, exercises and conditions that characterise the execution and can be used to foresee future consequences (Marks et al. 2016).

One example of a leading indicator is that of the near hit. This is characterised as an occurrence in which no harm to property or damage to an individual occurs, but which could have led to such consequences, given a slight change in time or position. The preference for this indicator is based on the fact that information can be gathered and analysed without the requirement that any damage to property or persons has occurred. Construction companies are required to report incidents which affect their business. Measurements derived from such reports, however, cannot reflect whether the consequences of hazard have been alleviated or the likelihood of harm has been reduced. Leading indicators are measures of procedures, exercises and conditions currently in place, and such indicators can anticipate future results. Unmitigated high-hazard circumstances can bring about significant or even lethal damage if they are permitted to persist. Linear causation models, such as loss-causation and domino models, argue that accidents are the final consequence of a succession of circumstances, and provide a sound motivation to gather and analyse close-hit information. It has also been found that, effectively, most significant accidents to individuals could have been prevented (Marks et al. 2016).

Construction safety is measurable from the behaviour patterns of the different stakeholders involved. This primarily relates to the behaviour patterns of workers as assessed via the ground theory method (Guo et al. 2015). Meanwhile, there are statistical analyses which specifically estimate the risks involved in a construction project. For instance, the risk assessments used in Saudi Arabian aviation construction utilise ANOVA tests to identify risks (Banhdadi and Kishk 2017). Alternatively, ontological modelling and document modelling are used to facilitate better measurement of the lack of knowledge present (Wang 2015). Through assessing the level of ignorance present concerning specific safety requirements, managers can attain a better awareness of the construction risks. To a certain extent, the above literature shows the potential of implementing automated approaches in the construction industry.

3 Automated Construction Tools

3.1 Robotics

Robots can be utilised for the execution of many different construction tasks, for example, the assembling of parts and the construction of the inside and outside skins of buildings. The operation of such robots is based on advanced control and detection technologies (Warszawski 1985). Some robots are designed to work in poor climatic conditions and harsh site environments (Skibniewski 1988). It has been demonstrated that robots could be used to mix and spread tarmac. Such applications use programmed guidance control with mechanical sensors and the calculation of a

Fig. 1 Bidirectional data
acquisition between virtual
and physical construction
(Akanmu et al. 2014)

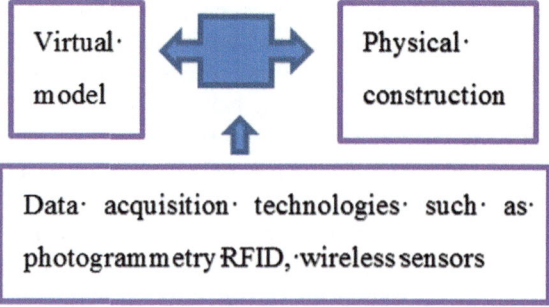

clearing velocity, and this latter leads to the controlled starting and stopping of all clearing capacities. Moreover, this work can be performed correctly using artificial vision and sensors. Also, a graphical remote-control framework allows a human operator to control the process progressively, continually altering the procedure. Window glass mounting or board settling can be undertaken to utilise a robot with a pneumatic actuator and a servo engine. Straightforward and independent robots can assemble two-dimensional structures and use pre-assembled modules as building pieces. Modules can do some data collection, empowering them to share long-range primary data and convey it to the robots (Elattar 2006).

3.2 Drone Photogrammetry

Photogrammetry strategies have been used for a long time together with CAD and virtual models of building structures—for architecture purposes. Photogrammetry utilises two images for measurement purposes along with a fundamental triangulation rule to calculate the areas of focus in three measurements. The relatively low cost and accessibility of digital cameras justify the use of photogrammetry-based methodologies for as-manufactured documentation. Cameras are less expensive, more compact and quicker regarding capturing relevant images, as contrasted with other information gathering devices. The captured images are later handled by software to obtain useful geometric information (Akanmu et al. 2014). With the popularity of various types of drones in Hong Kong, much of this photogrammetry work, in recent years, has been undertaken by relatively cheap drones instead of relatively expensive helicopters (Fig. 1).

3.3 Virtual Reality (VR) Simulation and Training

VR can create genuine work experience. This has been proposed as a viable form of safety training for the mining industry. VR can provide safety training exercises at any time, even when there is no experienced staff on hand to deliver the training. A personal computer is sufficient to provide the environment for VR training. Furthermore, it has been demonstrated that VR-based training is a useful technique for training workers in tower crane operation and safety procedures for ironworkers. Nevertheless, critics have commented that it is questionable whether VR frameworks can substitute for genuine experience. In Australia, some proposed VR training programmes have recognised areas in which formal training strategies and systems cannot replace the health and safety expertise acquired by experienced labourers. This situation motivates innovative thinking and has inspired specialists to think further about technologically based improvements to health and safety. VR systems have also been connected with the efforts of training specialists to act in and record safety training videos. They also allow users to collect safety and health data rapidly and effectively when the specialists need them. Thus, VR training is a successful technology for sharing and enhancing safety and health learning (Edirisinghe and Lingard 2016). In the US, VR allows various construction stakeholders to choose and select different building contents, whereas one Hong Kong construction company uses that to simulate the fall from a height to increase the safety awareness (Li 2018).

3.4 Modular Building Structure

Modular buildings are typically safer than conventional buildings regarding site construction because of the stable work area they impose. Regarding these, specialists can be fully knowledgeable about the dangers involved. Some of how safety is maintained is as follows: not working in tight spaces on site, performing off-site or on-ground assembly (as opposed to working from structures) and not working in severe weather conditions. With modular buildings, it is less painful to screen out dangerous activities, fewer energy needs are used on the construction site, and fewer local contractual workers and labourers are required in general. The modular building can mean assembling the entire building or just some parts off-site. The study discovered 125 mishaps associated with pre-assembled fabrication development with the most frequent types of the hazard being 'crack' and 'fall' hazards (Fard et al. 2015).

3.5 3D Printing

Many of us may take the view that the next step for modular buildings will be that of printing complete three-dimensional building structures. 3D printing has the potential to enable buildings to be constructed more quickly with lower costs, and with the further incorporation of building data into the development procedure. These potentialities mark it out as one of the new directions that the construction industry may take. Besides the fact that this technology could decrease the number of labourers required on sites (and so improve safety), it may also lead to a lessening in development times and an increase in engineering flexibility. Nonetheless, current 3D printing technology is not yet suited to large-scale items. These technologies are characterised by comprehensive configuration approaches and have an exceptionally constrained scope of materials that can be utilised (Perkins and Skitmore 2015). In Hong Kong, for example, some construction companies employ it for printing the 3D model before construction (Li 2018).

3.6 Building Information Modelling

In construction projects, there are various participants involved in different stages. Building Information Modelling (BIM) is a useful method that allows multiple stake-holders to cooperate together (Liu et al. 2017). BIM models as adjustable digital arte-facts contribute to their competence of functioning as apparatuses of exclusive design collaboration (Miettinen and Paavola 2018). Along with the use of data-collecting sensors, it can significantly reduce the probability of worksite accidents—because of the monitoring process used by the data acquisition system (Druley et al. 2016). In a more sophisticated examination of the consequences of these developments, it can be seen that the design of buildings, which is highly coherent with construction risks, can effectively be converted from a force-based design procedure into a probabilistic performance-based design procedure (Yarahmadi et al. 2017).

3.7 Intelligent Mixed Reality

We may also study the feasibility to train the computer via artificial intelligence to identify the hazard and inform the construction practitioners about the existence of risk by using artificial intelligence. With the help of the holographic mixed reality, we can approximate their location in real-life situations.

4 What Are the Leading Safety Indicators? A Glance at the Results from the Interviews

In this research, we use an information graphics (infographics) approach whereby we analyse and visualise the frequencies with which words are mentioned. The info-graphics which are often used in the mainstream media are the bar and line charts; these make data visually attractive and at the same time convey the message embed-ded in the information quickly (Li et al. 2015). These methods have been applied in previous research studies such as Li et al. (2016)—in that case, to analyse the objectives of competition law. In this paper, we adopt Tagul's technique to visualise the data as word clouds (Tagul 2016). The results from the interviews show that the most important reason leading to accidents in high-rise construction sites is seen to fall from height as indicated by the frequency of use (and therefore large size) of the associated words (Table 1).

5 Results from Court Cases Research

In the second part of the research, court cases which took place from 1991 to 2015 were reviewed. We extracted the factors that were considered by the judge to be the most important factors leading to construction accidents on sites. After that, utilising Tagul (2016) methodology, we used the infographics approach to study the factors that are more important. Stop words included 'contributory', 'provide', 'onto', 'defendant', 'mentioned', '1st, part', 'had' and 'common'. The results show that the most important reasons in the eyes of judges are related to negligence, equipment and legal issues. Unlike the practitioners, judges were more concerned with the indirect factors that lead to construction site accidents. Factors like bad weather were never the most important from their viewpoints (Fig. 2).

6 Case Study: Gammon Construction

In recent years, firms have been confronted with the difficulties involved in applying innovations, especially computer-based innovations, to their core working practices. The dissemination of development hypothesis is utilised to examine how advanced innovations diffuse crosswise over complex firms (Shibeika and Harty 2015). In the third part of the present study, we shed light on one construction company's innovative strategies for enhancing construction safety.

6.1 Curtain Wall—Robotic Installation

Gammon introduced the first curtain wall installation robot to Hong Kong; they did so to reduce the labour intensity of their operations. The use of robots reduces the quantity of manual operation required, and, in this case, relieved a serious shortage of curtain wall installers (the application of robots can reduce the number of installation workers by 25%), and it also enhances construction safety (Gammon Construction 2016b). Furthermore, to deal with the upcoming labour shortage due to the ageing of the workers involved, and to avoid on-site construction risks, robotics has become increasingly important (Gammon Construction 2016a). For instance, the company has introduced Exoskeleton and ZeroG Arm devices to assist workers in handling heavy loads/equipment. Specifically, Gammon provides an exoskeleton which can

Table 1 Factors that lead to construction accidents on high-rise as opposed to low-rise building sites

Five most important factors that lead to construction accidents on high-rise building sites	Five most important factors that lead to construction accidents on low-rise building sites
The most important factor	The most important factor

(continued)

Table 1 (continued)

Five most important factors that lead to construction accidents on high-rise building sites	Five most important factors that lead to construction accidents on low-rise building sites
The second most important factor	The second most important factor

(continued)

Table 1 (continued)

Five most important factors that lead to construction accidents on high-rise building sites	Five most important factors that lead to construction accidents on low-rise building sites
The third most important factor	The third most important factor

(continued)

Table 1 (continued)

Five most important factors that lead to construction accidents on high-rise building sites	Five most important factors that lead to construction accidents on low-rise building sites
The fourth most important factor	The fourth most important factor

(continued)

Table 1 (continued)

Five most important factors that lead to construction accidents on high-rise building sites	Five most important factors that lead to construction accidents on low-rise building sites
The fifth most important factor	The fifth most important factor

Fig. 2 Judges' viewpoints about all types of construction work on sites

automatically sense the workers' movement, then provide back support for the workers so that they can lift heavy objects. This can reduce the risk of back injuries very effectively (Gammon Construction 2016a). At the same time, the ZeroG Arm can help workers to manoeuvre heavy equipment. The energy loss by workers can be greatly reduced as this last device provides an experience of near weightlessness about the load/heavy equipment, and so allows operations to be performed more quickly and easily (Gammon Construction 2016a).

6.2 Virtual Reality Safety Training

Online gaming is a common trend nowadays, especially among the younger generation. Gammon is the forerunner in incorporating virtual reality (VR) into safety training. Construction sites are simulated in 2D and 3D game environments to enable trainees to experience different scenarios in virtual construction sites. Safety Manager Kwok Wai-yin comments that training, as such, is more effective and convincing than lectures. It can draw the trainees' attention to specific issues, stimulate their responses and enhance mutual communication. It changes the workers' mode of thinking, boosts site safety and enables the firm to approach closer to the goal of zero accidents on sites (Gammon Construction 2016b).

6.3   i-Core Development

It often requires a great deal of work to track down the location of materials when they are in transit from where they are produced to the location at which they are to be installed on a site. i-Core is a standalone, extendable and programmable electronic circuit with GPS, GSM, a thermometer, Bluetooth, a gyroscope and a barometer, all integrated into one i-Core chip. Equipped with GSM/GPS technologies, Gammon's i-Core provides a logistics-management solution for expensive construction materials such as chiller plants and precast concrete products. It can track real-time locations and share location information via an online database. It enhances efficiency, productivity and saves human resources in the tracking process (Gammon Construction 2016b).

6.4   Modular Construction

For the Tuen Mun–Chek Lap Kok Link—Southern Connection Viaduct Section venture, the company augmented its use of precast items for the marine viaduct components. An example of such a component was the precast heap top shells. The shells are fabricated off-site and are then simply put on top of the perpetual heaps and dewatered to bolster the low heap top's development, without the need for cofferdams or marine falsework. Thus, the number of rebar fixers, welders, woodworkers, riggers and jumpers could be diminished hugely. The precast approach likewise decreases the quantities of materials used, the amount of work needed at the structure itself, and minimises the likelihood of unintentionally dropping cement into the ocean (Gammon 2015). Precast heap top shells saved 40,000 working days and had a vital advantage regarding worker well-being since specialists needed to work for less time at the location itself (Gammon 2015).

6.5   Drone Photogrammetry for Surveying

The imaginative utilisation of automata and drone photogrammetry provides supervisory work with more flexible resources and results in the use of fewer area surveyors, and also lowers costs. Whilst the expense of current reviewing work per square metre is HK$2, the expense of photogrammetry per square metre is HK$0.1 (Gammon 2015).

6.6 BIM Technology and 3D Scanning

In 2013, Gammon advocated the use of Building Information Modelling (BIM). BIM is a 3D-model-based process which facilitates the collaboration of the participants in a construction project. It helps the management of the project to draft detailed project designs and to plan. Even the initial plan is much better than a traditional plan would be, in that the considerations are well rounded so that the plan is optimised to deal with actual needs. Throughout the construction process, resource consumption is under BIM supervision. BIM suggests the way to estimate and allocate resources to be efficient, reduce costs and to represent less of a burden to the environment. Thus, after adopting BIM, construction projects can be simulated digitally to achieve higher efficiency and lower waste production (Gammon Construction 2013). The famous buildings for which BIM was adopted throughout their construction include Hysan Place, One Island East and Lee Garden Three. Moreover, with assistance from BIM, Gammon achieved a step forward towards adopting 3D printing and scanning technologies. Through 3D technologies, the construction projects will be more efficient and accurate from the design stage onwards.

6.7 AI Chatbot for Filling in the Accident Report and Smart Concrete Sensor

AI Chatbot is used on sites which can help the contractor fill the safety investigation report. Traditionally, a report filled by hand can hardly be standardised. Nevertheless, the AI chatbot automatically standardise the items such as 3/F, 3rd floor, 3f to one single representation. Once the safety officers have answered all the questions sent from the chatbot, a safety investigation report can be generated automatically.

The smart sensor utilises the Internet of Things to work. Once it detects that the concrete reaches a certain level of heat, it will stop the heating process, substantially reduce the workers' work and thus reduce the accident rates.

7 Discussion and Conclusion

The results of the interviews showed that the practitioners considered that falls from heights are the most important factors which lead to accidents on sites, followed by bad weather and machine faults. On the other hand, the top three factors for low-rise buildings' construction related to the lack of important safety equipment, training, etc. Hence, the idea 'lack' is spelt out explicitly in the diagram. The second and third most important factors are machine faults and lousy weather. As regards the judges' perceptions, negligence is considered to be the most critical factor.

Can innovation involving the various information technologies available nip construction accidents in the bud? The introduction of robotics, prefabricated/precast building structures, i-core with GPS, aerial photogrammetry and 3D printing implies that the number of workers on site can be reduced, leading to a possible reduction in construction accidents. Also, VR training can reduce the likelihood of mishaps occurring due to a lack of training and/or knowledge concerning construction activities on sites.

Acknowledgements An earlier version of the chapter was included in the conference proceeding: Li, Rita Yi Man and Leung, Tat Ho (2017) Construction safety in Hong Kong: an automated approach, 34th International Symposium on Automation and Robotics in Construction, Taipei, Taiwan, 28 June 1 July 2017. We would also like to thank an interviewee from Gammon in providing their latest development in smart concrete sensor and AI Chatbot in 2018.

References

Akanmu, A., C. Anumba, and J. Messner. 2014. Critical review of approaches to integrating virtual models and physical construction. *International Journal of Construction Management* 14 (4): 267–282.

Alarcón, L.F., D. Acuña, S. Diethelm, and E. Pellicer. 2016. Strategies for improving safety performance in construction firms. *Accident Analysis and Prevention* 94: 107–118.

Banhdadi, A., and M. Kishk. 2017. Assessment of risks associated with Saudi aviation construction projects and of the risk's importance. *International Journal of Innovation, Management and Technology* 8 (2): 123–130.

Bong, S., R. Rameezdeen, J. Zuo, R.Y.M. Li, and G. Ye. 2015. The designer's role in workplace health and safety in construction industry: Post-harmonized regulations in South Australia. *International Journal of Construction Management* 15 (4): 276–287.

Druley, K., T. Musick, and S. Trotto. 2016. Researcher explores how to make temporary structures on the construction site safer. *Safety & Heath* 193 (3): 35.

Edirisinghe, R., and H. Lingard. 2016. Exploring the potential for the use of video to communicate safety information to construction workers: Case studies of organizational use. *Construction Management and Economics* 34 (6): 366–376.

Edrei, T., and S. Isaac. 2016, forthcoming. Construction site safety control with medium-accuracy location data. *Journal of Civil Engineering and Management*.

Elattar, S.M.S. 2006. Automation and robotics in construction: Opportunities and challenges. *Emirates Journal for Engineering Research* 13 (2): 21–26.

Fard, M. M., S. A. Terouhid, C. J. Kibert, and H. Hakim. 2015. Safety concerns related to modular/prefabricated building construction. *International Journal of Injury Control and Safety Promotion*.

Gammon. 2015. *Gammon construction harness technology novelty in moderating labour intensity* [cited 7 August 2016]. Available from http://www.gammonconstruction.com/en/html/press/press-bd530331cdef40989c4520fafb1aa13f.html.

Gammon Construction. 2013. *Gammon advocated the application of BIM and was awarded the ISO 14064 verification statement. Gammon Construction* [cited 6 March 2019]. Available from https://www.gammonconstruction.com/en/html/press/press-07b0f78d171b41d08265979a0aa8c2c7.html.

Gammon Construction. 2016a. *Build-to-order full modular and robotic construction. Gammon construction* [cited 6 March 2019]. Available from https://www.gammonconstruction.com/en/html/press/press-ba2c50896ee0486497849c742c7ac045.html.

Gammon Construction. 2016b. *Gammon construction extends innovative technology efforts to optimize construction—Virtual reality safety training, GPS tracking, robotic construction* [cited 7 August 2016]. Available from http://www.gammonconstruction.com/en/html/press/press-6b9ebe23ad3a426b847f03c4e5fa7031.html.

Geoff, E. 2016. Major construction projects at airports: Client leadership of health and safety. *Journal of Airport Management* 10 (2): 131–137.

Guo, B.H.W., T.W. Yiu, and V.A. Gonzalez. 2015. Identifying behaviour patterns of construction safety using system archetypes. *Accident Analysis and Prevention* 80: 125–141.

Hoła, B., M. Sawicki, and M. Skibniewski. 2015. An it model of a knowledge map which supports management in small and medium-sized companies using selected polish construction enterprises as an example. *Journal of Civil Engineering and Management* 21 (8): 1014–1026.

Li, R.Y.M. 2015. *Construction safety and waste management: An economic analysis*. Germany: Springer.

Li, R.Y.M. 2018. *An economic analysis on automated construction safety: Internet of things, artificial intelligence and 3D printing*. Singapore: Springer.

Li, R.Y.M., H.C.Y. Li, K. Mak Cho, and P.K. Chan. 2016. Rationales for the implementation of competition law in EU, the US and Asia: Content analysis and data visualization approach. *Asian Journal of Law and Economics* 7 (1): 63–100.

Li, R.Y.M., and S.W. Poon. 2013. *Construction safety*. Berlin: Springer.

Li, Z., S. Carberry, H. Fang, K.F. McCoy, K. Petersona, and M. Stagitis. 2015. A novel methodology for retrieving infographics utilizing structure and message content. *Data & Knowledge Engineering* 100: 191–210.

Liu, Y., S. Van Nederveen, and M. Hertogh. 2017. Understanding effects of BIM on collaborative design and construction: An empirical study in China. *International Journal of Project Management* 35 (4): 686–698.

Marks, E., I.G. Awolusi, and B. McKay. 2016. Near-hit reporting reducing construction industry injuries. *Professional Safety* 61 (5): 56–62.

Miettinen, R., and S. Paavola. 2018. Reconceptualizing object construction: The dynamics of Building Information Modelling in construction design. *Information Systems Journal* 28 (3): 516–531.

Mroszczyk, J.W. 2015. Improving construction safety. *Professional Safety* 60 (6): 55–68.

Perkins, I., and M. Skitmore. 2015. Three-dimensional printing in the construction industry: A review. *Journal: International Journal of Construction Management* 15 (1): 1–9.

Shibeika, A., and C. Harty. 2015. Diffusion of digital innovation in construction: A case study of a UK engineering firm. *Construction Management and Economics* 33 (5/6): 453–466.

Skibniewski, M.J. 1988. *Robotics in civil engineering*. New York: Van Nostrand Reinhold Pub.

Skibniewski, M.J. 2014. Research trends in information technology applications in construction safety engineering and management. *Frontiers of Engineering Management* 1 (3): 246–259.

Tagul. 2016. *Tagul—Word cloud art* [cited 9 August 2016]. Available from https://tagul.com.

Wang, H.H. 2015. Semi-automated identification of construction safety requirements using ontological and document modeling techniques. *Canadian Journal of Civil Engineering* 42 (10): 756–765.

Warszawski, A. 1985. Economic implications of robotics in building. *Building and Environment* 20 (2): 73–81.

Yarahmadi, H., M. Miri, and M. Rakhshanimehr. 2017. A methodology to determine the response modification factor for probabilistic performance-based design. *Bulletin of Earthquake Engineering* 15 (4): 1739–1769.

Chapter 10
The Open Building: A Review of Its Economics, Environment and Related Social and Safety Issues

Abstract In this chapter, we discuss the concept of an open building as it is described in many different electronic databases. The term open building refers to a building which can be used flexibly according to a user's needs, without demolition. The use of open buildings is possibly one of the most effective means of creating a sustainable built environment. We also study the viewpoints of academic researchers, and so look at to previous research papers on the concept of open buildings.

Keywords Sustainable development · Open building · Knowledge sharing

1 Introduction

As the ever-changing needs of society often lead to building obsolescence, modern builders and architects have been looking at many different ways to solve this problem. A somewhat new form of architectural design, which is often designated by the term 'open building', has been adopted for some recent building projects. Constructing open buildings has been defined as using the principles of ordering and combining subsystems (Jaillon and Poon 2014), so that no demolition is necessary for the building structure as a whole when alterations or additions are required later, once the building has become occupied. Open buildings enable residents to change their dwellings and layouts according to their changing needs and preferences without affecting their neighbours. The concept stresses the importance of customisability and flexibility for adaptation.

The concept of open buildings has been widely used for both residential and non-residential buildings (Kendall 1999). It separates the building structures into fixed and changeable elements (Wong 2010) and provides useful information for customising the apparent unsystematic process of architectural design. It thus has become one of the significant fields of knowledge and experience in construction (Bluyssen et al. 2010). The idea of open residential buildings also represents a significant paradigm shift in the housing (Kendall 1999).

© Springer Nature Singapore Pte Ltd. 2019
R. Y. M. Li, *Construction Safety Informatics*,
https://doi.org/10.1007/978-981-13-5761-9_10

The principles of open buildings provide a systematic approach to building construction (Chien and Wang 2014). Supports and infills are the two primary mechanisms which provide a systematic way to link, vary or customise the building element according to the residents' needs. This highly technical building method applies straightforward building materials so that the flexibility can be sustained (Schneider and Till 2005).

2   Technical Aspects of Open Buildings' Energy Conservation and Thermal Transfer

For natural convection, the scales and dimensional variables are as follows:

$$(X, Y) = (x, y)/H, (U, V) = (u, v)/U_0, T = (t - t_{\text{atm}})/\Delta t, \text{ Pts}$$
$$= P/pU_0^2 \tag{1}$$

The two-dimensional equations which describe the idea of continuity, energy and momentum transport can be expressed accordingly as follows:

$$\frac{\partial U}{\partial X} + \frac{\partial V}{\partial Y} = 0 \tag{2}$$

The X-momentum equation will be

$$\frac{\partial UU}{\partial X} + \frac{\partial VU}{\partial Y} = -\frac{\partial P}{\partial X} + \frac{\partial}{\partial X}\left(\text{Ptsr}\frac{\partial U}{\partial X}\right) + \frac{\partial}{\partial Y}\left(\text{Ptsr}\frac{\partial U}{\partial Y}\right) \tag{3}$$

The Y-momentum equation will be

$$\frac{\partial UV}{\partial X} + \frac{\partial VV}{\partial Y} = -\frac{\partial P}{\partial Y} + \frac{\partial}{\partial X}\left(\text{Ptsr}\frac{\partial V}{\partial X}\right) + \frac{\partial}{\partial Y}\left(\text{Ptsr}\frac{\partial U}{\partial Y}\right) + Ra\text{Ptsr}T \tag{4}$$

And the energy conservation equation is

$$\frac{\partial UT}{\partial X} + \frac{\partial VT}{\partial Y} = -\frac{\partial^2 T}{\partial X^2} + \frac{T\partial^2}{\partial Y^2} \tag{5}$$

where n and a, respectively, represent the kinematic viscosity and the thermal diffusivity and b is the thermal volumetric expansion coefficient.

The heat transfer rate is calculated as a superposition of convection and conduction. Thus, it is a spatial generalisation of the Nusselt number concept, in the sense that it describes the magnitude and direction of the heat transfer rate through any surface that can be imagined inserted into the convective medium. Regarding the continuity and energy conservation equations in the fluid domain, the dimensionless stream function Φ and the heat function q is defined, respectively, as follows:

Stream function:

$$\partial \Phi / \partial Y = U, \quad -\partial \Phi / \partial X = V \tag{6}$$

Heat function:

$$\partial \Phi / \partial Y = \frac{\mathrm{Pr}\, UT - \partial T}{\partial X} = \mathrm{PrVT} - \frac{\partial T}{\partial Y} \tag{7}$$

These transport functions become dimensionless when

$$\Phi = \Phi^* / p U_0 H \theta = \theta^* / \Delta t \mathbb{C} \tag{8}$$

The convective heat transfer rate of the heating source and ventilation performance should be evaluated—i.e. the global Nusselt number (Nu) over the heated wall and the overall exchange rate of mass (M) through the inlet opening would be emphasised. Due to the application of non-slip boundary conditions, the fluid flow subsides in the surface of walls. Thus, Nu is obtained by integrating along the heated wall as given below:

$$\mathrm{Nu} \int_{L_{\text{heat}}} \frac{\partial T}{\partial n} \mathrm{d}L \tag{9}$$

where L_{heat} represents the heated surface length and the mass exchange rate (M) through the vented opening will be

$$M \int_{L_{\text{in}}} V_n \mathrm{d}L \tag{10}$$

where L_{in} is the opening inlet's length; the opening is distributed in the right side's lower region, and the fluid flow is governed by Eqs. (2–5) (Table 1).

3 Materials and Methods

In addition to demonstrating the sustainability aspects of open buildings, we include a scholarly review as well as a description of a new invention that has obtained a patent. Several databases maintained by the University of Science and Technology are used in this study, these include 'Science Direct', 'Emerald Emerging Markets Case Studies', 'Proquest' and 'EBSCOhost Research Databases'. After examining these, we then

Table 1 Summary of the symbols used when calculating an open buildings' energy conservation and thermal transfer (Zhang et al. 2017)

		Greek symbols		
$\mathbb{C}$	Thermal conductivity (W/m k)		AR	Aspect ratio of the enclosure g gravitational acceleration (m/s^2)
α	Thermal diffusivity (m^2/s)		H	Height of the rectangle, square (m)
β	Volumetric expansion coefficient		M	Dimensionless mass flow rate n unit outward normal vector
Δ	Difference in value		p	Pressure (N/m^2)
ν	Kinematic viscosity (m^2/s)		P	Dimensionless pressure
Pts	Density (kg/m^3)		Pr	Prandtl number
$\square$	Generic intensive variable		Ra	Rayleigh number
Φ	Stream function		T	Temperature (K)
ζ	Heat function		T	Dimensionless temperature
	Subscripts		u, v	Velocity components in x, y (m/s)
atm	Ambient environmental parameters		U, V	Dimensionless velocity
high (low)	Higher (lower) value (out) into (out of) cavity		U_0	Reference velocity scale (m/s)
nb	Neighbouring nodal points		W	Width of the rectangle, square (m)
	Superscripts		x, y	Cartesian coordinates
*	Dimensional		(m)	X, Y dimensionless coordinates

searched for 'open building' via Google Patent to uncover the current inventions which are related to open building sustainable. The results are then analysed via systematic content analysis approach.

Content analysis is then used to analyse the results, similar information is grouped into related categories to create an objective and systematic criterion for writing text for investigating the symbolic content of communication (Li 2013, 2015).

4   Results of the Open Building Research

We found that the number of relevant publications in the academic databases was limited. Table 2 shows the related articles found in the various databases by using the keywords: open building and 'open building'. From the table, we can see that most of the related journal articles were found in 'Proquest'. Table 3 shows the paragraphs related to the definitions of the open building that can be found in the journal articles (Table 4).

Table 2 Number of articles that are related to open buildings

	Keywords	Number of articles	Number of relevant articles
Science Direct	Open building	338,578	2
Science Direct	'Open building'	217	2
Emerald	Open building	78,675	0
Emerald	'Open building'	31	1
Proquest	Open building	543,677	2
Proquest	'Open building'	93	6
EBSCO	Open building	4472	0
EBSCO	'Open building'	14	1

4.1 Economic, Social and Environmental Aspects of Open Buildings

In economic terms, some scholars believe that 'open building' could minimise the industrial waste produced because of construction activities (Nadim and Goulding 2011; Schneider and Till 2005; Kendall 1999; Till and Schneider 2005; Jaillon and Poon 2014; Cao et al. 2015). The characteristic of flexibility reduces the expense involved with reassigning a building's use, and this implies that such buildings are much more economical in the long run (Schneider and Till 2005). Whilst cabinets, finishes and furniture are all components which can be adjusted or disassembled and so can be changed anytime (Ness et al. 2014), open building allows, in addition, changes to an entire environment or structure as required, when the usage of the building changes, without any demolition of the building is necessary (Jaillon and Poon 2014). Thus, in this regard at least, society's resources can be recycled, and therefore resource efficiency and resource productivity can be enhanced.

In social terms, these components enable design adaptability, fast on-site installation and future changeability (Chien and Wang 2014). Thus, an 'open building' adapts to different usages and differing users' preferences over time (Schneider and Till 2005). Thanks to the ability to disassemble parts of the structure dissemble, improvements in living conditions can be separately or jointly achieved, and new functional demands and user-defined adjustments can be catered for (Cao et al. 2015; Gosling et al. 2013). In the open building approach, a multi-unit building is constructed so that a variety of the different occupants' preferences can be satisfied and this costs the developer no more than would make all the units the same without any such creative input (Kendall 1999). Besides, households can share a base building of higher quality with a longer life. The base architecture provides the possibility of variation at the level of every individual dwelling unit. Public libraries which adopt the 'open building approach' attain social and cultural sustainability as they are open buildings physically, carry messages not only in text and digital media, but also the architectural design, for example, Oslo library (Edwards 2011).

Table 3 Definition of open building

Author	Definition
Timothy (2011)	Open building stresses the need for customizability and flexible systems which anticipate the needs for adaptation
Carmon (2002)	The open building system means 'buildings that enable residents to shape their dwellings and change layouts according to the changing needs and preferences without interfering with the neighbours'
Chien and Wang (2014)	The open building principles offer an orderly approach for building manufactures. The supports consist of long-lasting and common building services. An infill system involves a pre-packaged, joined set of products and a custom prefabricated off-site for a dwelling. The process allows incremental changes and enhancements of the existing buildings
	The concept of open building divides a building into supports and infills. This division links the building elements systematically and permits changes or customisation according to the needs of the occupants. Van Randen and Habraken established the Matura infill system with the matrix tile and the baseboard profile. The matrix tile system is designed according to a 10 cm × 20 cm grid concerning the component position and its relationship to other building elements. The baseboard that consists of a track component serves as the basis for interior partitions, provides design flexibility, rapid on-site fitting and subsequent variations
Schneider and Till (2005)	The best-known constructional method to ease flexibility in housing and enable users' participation
Wong (2010)	The open building approach separates the building into fixed and changeable features. The adaptability of a building depends on how well the fixed elements are organised to provide rooms for variations in the elements. This property is distinct from flexibility as the provision for space used for various purposes without physical alterations. While the adaptability of a building designates its physical and morphological characteristics, the flexibility of a building refers to its functional features. The former attribute involves the alteration in the physical characteristics of the building such as dimensions, form, openings, etc., while the latter ones do not. The characteristics of flexibility and adaptability of a building link to the dimensional and functional requirements
Jaillon and Poon (2014)	The open building system design permits fitting out alterations. The temporal hierarchy of building element displays that building structure has long lives. Some may have proposed the parting of the building base from the 'infill', i.e. the fitting out throughout the design and building stages of the housing structure to increase the contribution of the occupant in the structural design process

(continued)

Table 3 (continued)

Author	Definition
Gaosling (2013)	The open building movement separates the base building from its interior fit-out, which can be installed or changed with a minimum interface problem. Thus, it enables the building to adjust according to the users' preferences over time
Koebel and Edwards (1999)	Open building is an approach to design, architecture and long-term buildings adaptation. The open building approach confines rigid structural and mechanical topographies to the base building, with all other spaces planned for flexible and variable 'fit-out' for diverse dwellers over time. Open building improves sustainability by evading premature obsolescence, costly renovation and retrofitting

Environmentally speaking, assemblies and components can be disassembled, reconfigured, reused and re-fabricated (Ness et al. 2014). On a shorter time frame, cabinets, finishes and furniture can be adjusted (Kendall 1999). Therefore, the wastes caused by industrial manufacture can be reduced. The 'fit-out' under the 'open building approach' should be able to be installed and altered with a minimum of issues, enabling the building to adapt to users' preferences over time (Gosling et al. 2013).

Notwithstanding all this, some scholars take the view that open buildings cannot provide achiever sustainability. Table 5 shows the relevant articles found across some different databases by using two crucial different word sequences—open building and sustainability and 'open building' and sustainability. From this table, we can see that most of the related journal articles can be found in the 'Proquest' database. It seems that the authors of most of these articles are convinced that 'open buildings' can indeed attain certain levels of sustainability—i.e. be economically, socially and environmentally friendly.

4.2 The Construction and Building Safety of Open Buildings: A Patent Review

The invention considered here is that of a reusable safety rail which can improve environmental sustainability about the open floors of a building under construction. The track comprises an outer stanchion rigidly and detachably secured to an inner stanchion; the inner stanchion is secured to a floor portion of the building under construction and multiple guard members can be removed, ably secured to the outer stanchion (Christoffer and Stawychny 2013).

Safety rails may be utilised to keep labourers from tumbling from open floors of structures under development. This kind of track is by and large made of wood which is anchored together with nails. The wooden safety rails are, in the end, removed and the material from which the wooden safety rail is made is thrown out from the installation zone. This way of assembling and dismantling security rails is labour

Table 4 Articles that are related to sustainable open building

Name of the journal database	Keywords	Number of articles	Number of relevant articles
Science Direct	Open building and sustainable development	47,574	0
Science Direct	'Open building' and sustainable development	35	2
Science Direct	Open building and sustainability	26,687	0
Science Direct	'Open building' and sustainability	21	4
Emerald	Open building and sustainable development	26,782	0
Emerald	'Open building' and sustainable development	15	1
Emerald	Open building and sustainability	14,565	0
Emerald	'Open building' and sustainability	11	0
Proquest	Open building and sustainable development	42,041	0
Proquest	'Open building' and sustainable development	23	3
Proquest	Open building and sustainability	31,408	0
Proquest	'Open building' and sustainability	22	3
EBSCO	Open building and sustainable development	49	0
EBSCO	'Open building' and sustainable development	2	2
EBSCO	Open building and sustainability	74	0
EBSCO	'Open building' and sustainability	1	1

Table 5 Showing the degree of sustainability of the open building

Academic article	Economic	Social	Environment
Factors affecting open building implementation in high-density mass housing design in Hong Kong		✓	✓
Smart partition system—a room-level support system for integrating smart technologies into existing buildings		✓	
Off-site production: a model for building down barriers	✓	✓	✓
Flexible housing: opportunities and limits	✓	✓	
Open building: an approach to sustainable architecture	✓	✓	✓
Sustainability as a driving force in contemporary library design		✓	
Flexible housing: the means to the end	✓	✓	
From the guest editors	✓	✓	✓
Smart steel: new paradigms for the reuse of steel enabled by digital tracking and modelling	✓		✓
The study of factors that inhibit the promotion of the SI housing system in China	✓	✓	✓
Life cycle design and prefabrication in buildings: a review and case studies in Hong Kong	✓	✓	✓
Adaptable buildings: a systems approach		✓	

intensive, tedious and costly. After the work is finished regarding the individual open floor in question, the wooden safety rails are torn apart and disposed of, and the finished floor is then encased. New timber is then required for making safety rails for the next open floor of the building. The wood which has been disposed of will usually end up in a landfill. The new wood for the following safety rail must be estimated for, cut and fitted in the same labour intensive, tedious way. The new timber is then disposed of when the work is finished for this next floor. The procedure is repeated until the building is completed (Christoffer and Stawychny 2013).

Thus, there is often a need for a metal safety rail that is reusable and can be rapidly dismantled and gathered up. This particular patented type of rail disassembles into two separate types of unit to keep coordinating components from being lost or damaged. It is also more stable, safe and secure than any wooden safety rail would be. Furthermore, such a metal security rail must meet all specifications set out by safety offices (Christoffer and Stawychny 2013).

5   Conclusion

Given the advantages of open building, it is odd that this type of building is not accessible, and not in evidence in many cities. One of the reasons for this is that considerable efforts are required to design such a building (Wong 2010). Not only must it serve its present functions well, but it also must be adaptable and flexible enough to serve the future users' needs. Possibilities for reconfiguration need to be considered before the building is constructed—in the design stage. Moreover, such 'open buildings' will not eliminate the need for the periodic retrofitting or renovation of housing. Obsolescence of the building must be solved at some time through renovation, and improvements in technology will play a part in this. Additionally, some materials and products used today might be revealed as health hazards in the future. It is crystal clear that this was in the past the case with lead-based paint which represents a significant health hazard in older housing, especially this occurs in units used by low-income households.

References

Bluyssen, P., M. Oostra, and H. Böhms. 2010. A top-down system engineering approach as an alternative to the traditional over-the-bench methodology for the design of a building. *Intelligent Buildings International* 2 (2): 98–115.

Cao, X., Z. Li, and S. Liu. 2015. Study on factors that inhibit the promotion of SI housing system in China. *Energy and Buildings* 88: 384–394.

Chien, S.-F., and H.-J. Wang. 2014. Smart partition system—A room level support system for integrating smart technologies into existing buildings. *Frontiers of Architectural Research* 3 (4): 376–385.

Christoffer, A.C., and S.M. Stawychny. 2013. *Metal safety for open floors of a building under construction.* In U. S. Patent (ed.), vol. US 8.424,851 B2. US.

Edwards, B. 2011. Sustainability as a driving force in contemporary library design. *Library Trends* 60 (1): 190–214.

Gosling, J., P. Sassi, M. Naim, and R. Lark. 2013. Adaptable buildings: A systems approach. *Sustainable Cities and Society* 7: 44–51.

Jaillon, L., and C.S. Poon. 2014. Life cycle design and prefabrication in buildings: A review and case studies in Hong Kong. *Automation in Construction* 39: 195–202.

Kendall, S. 1999. Open building: An approach to sustainable architecture. *Journal of Urban Technology* 6 (3).

Li, R.Y.M. 2013. The usage of automation system in smart home to provide a sustainable indoor environment: A content analysis in web 1.0. *International Journal of Smart Home* 7 (4): 47–60.

Li, R.Y.M. 2015. *Construction safety and waste management.* Berlin: Springer.

Nadim, W., and J. Goulding. 2011. Offsite production: A model for building down barriers. *Engineering, Construction and Architectural Management* 18 (1): 82–101.

Ness, D., J. Swift, D. Ranasinghe, K. Xing, and V. Soebarto. 2014. Smart steel: New paradigms for the reuse of steel enabled by digital tracking and modelling. *Journal of Cleaner Production.*

Schneider, T., and J. Till. 2005. Flexible housing: Opportunities and limits. *Arq: Architectural Research Quarterly* 9 (02).

Till, J., and T. Schneider. 2005. Flexible housing: The means to the end. *Arq: Architectural Research Quarterly* 9 (3–4).

Wong, J.F. 2010. Factors affecting open building implementation in high density mass housing design in Hong Kong. *Habitat International* 34 (2): 174–182.

Zhang, J.H., D.D. Zhang, D. Liu, F.Y. Zhao, Y. Li, and H.Q. Wang. 2017. Free vent boundary conditions for thermal buoyancy driven laminar flows inside open building enclosures. *Building and Environment* 111: 10–23.

Chapter 11
The Epistemology of the Causes that Lead to Accidents on Sites: The Institutional and the Austrian Economics Approach

Abstract In this chapter, we aim to study the judges' viewpoints, i.e. the viewpoints of those who must consider all the various items of evidence before they can construct a judgment bearing in mind the positions of all the various stakeholders. The judges may be relied upon, perhaps, to maintain an objective understanding of the causes of construction accidents. After this, various practitioners' thoughts on what causes construction accidents are looked at. Their viewpoints are studied under the framework of Institutional and Austrian economics.

Keywords Epistemology · Institutional economics · Austrian economics

1 Introduction

Safety is one of the most essential issues in the construction industry. There is, however, no consensus concerning the factors which lead to construction accidents. Previous discussions in the literature often seem to have no common ground. By collecting 216 construction workers' cases from 29 construction sites, including 36 critical incidents of hyperthermia which were reported, and interviewing 96 managers, having 37 different job titles, of both client and contractor organisations, Rowlinson and Jia (2015)'s inductive approach suggested that construction accidents are related to heat stress on sites. They demonstrated the effects of the environment and the individual's physical activity on construction accidents. Their research provided evidence that there are eight levels at which factors involved in construction accidents can exist: ecosystem, society, industry, permanent organisation, project organisation, team, job unit and the individual level. Salminen et al. (1990) argued that organisational factors account, to a large extent, for construction accident causalities—according to workers' interviews. It was advised that organisational factors should be included in prevention practices.

By conducting focus group research and studies of 100 individual construction accidents, Haslam et al. (2005) conclude that 70% of accidents happen due to the workers involved or their team. Other factors include workplace issues, problematic equipment, the suitability and condition of materials, and deficiencies in risk management (84%); these are the significant factors that lead to construction accidents. After examining a model representing causal influences, Haslam et al. (2005) pointed out that it is necessary to pay more attention to originating influences to have sustained construction safety. Similarly, Gambatese et al. (2008) provided evidence that design professionals also play an important role in construction safety—by reviewing 224 fatality cases and judging for each whether the design was a factor in the incident or not.

Collecting data from construction accident reports produced in Spain between 2003 and 2008, Arquillos et al. (2012) concluded that the causes of construction accidents in Spain were mainly related to age, size of company, length of service, location of accident, day of the week, days of absence, deviation from accepted practice and climatic zones. Li (2018) and Li and Poon (2013) proposed that lack of knowledge, human error, hectic schedule, weather and piece rate are some of the factors that lead to accidents on construction sites.

Brogmus (2007) and Amiri et al. (2014) found that accidents usually happen on Mondays. Similarly, Yilmaz (2015) analysed the statistical data collected from a construction site in Istanbul—Turkey, which included data on 200 occupational accidents that occurred between February 2012 and November 2013. And he concluded that most of the construction accidents occurred on Monday, in summer and between the 1st to the 6th working hours. He hypothesised that the reasons were the lack of motivation after the weekend holiday.

2   The Division of Labour Implies a Division of Knowledge

Why do we observe that different stakeholders usually hold different views with regards to construction accidents? In this chapter, we shed light on the construction safety knowledge held by those who have come across accidents before and that of judges involved; our investigation will be based on knowledge analysis.

Scientific knowledge is considered a public good and the cause of positive externalities. The non-scarce nature of experience, once it has been discovered, implies that its use is unlimited and its non-excludability means that its use is limited. In the absence of excludability, there can be insufficient investment in knowledge, and this needs to be remedied via government subsidies or intellectual property rights. The idea of Jacobs spillover suggests that once technological knowledge is generated, the dissemination of such knowledge creates new knowledge. While efficient dissipation depends on knowledge, and the efficient use of knowledge depends on the existing technological knowledge (Leppälä 2015).

Market knowledge can be learnt via trial and error, automatic discovery or via thoughtful search. It can be dispersed from one individual to another and can ultimately be used in decision-making. Many of the unanswered economic problems are related to the questions of knowledge. The required knowledge is usually held by individuals, scattered across different sections of society and so is not likely to aggregate. Despite Adam Smith's division of labour being beneficial to our community, this inevitably implies a division of knowledge: each of us knows different things (Leppälä 2015) due to this different people's views will be incompatible with different cognition focused on one process (Hayek 1945) and direct access to others' local knowledge is costly (Leppälä 2015).

Knowledge can be shared by observation or communication and can only be absorbed individually. No collective mind contains all knowledge. While the use of ICT greatly enhances communication, most explicit knowledge alludes to tacit knowledge. Knowledge spillovers are local public goods that generate positive externalities. They provide hints on how innovative know-how is created, diffused, adapted and combined by individuals. Hence, knowledge externalities are the engine of growth as well as a symptom of a market failure (Leppälä 2015).

While labour mobility between firms induces knowledge spillovers, the labour market is effective in externalities internalisation. Tacit knowledge refers to different kinds of skills. For example, riding a bicycle is an example that is usually given. It is that element of our understanding that we are unable to communicate and accounts for the reason why some experience is not immediately communicated and commonly shared—as observers cannot understand it, and it has no informative content. The idea of tacit knowledge explains why some of the knowledge in society resists becoming widely disseminated. Other knowledge is not shared because it is not economical to do so. Moreover, even where it is communicated, this does not imply that knowledge is transferred since incentives matter (Leppälä 2015). Nevertheless, not all the knowledge that an individual acquires will be useful (Hayek 1937). Knowledge is the outcome of unconscious experiments via trial and error (Vanberg 2014).

3 Epistemic Knowledge and Truth

Knowledge is the accumulation of patterns and regularities that result in organised explanations of their environments (Leppälä 2015). It is incidental to anthropoid action. It arises from observations related to activities or in-depth knowledge seeking (Li and Poon 2011). People have more sophisticated epistemology if they are aware of the significant role of ideas in the knowledge acquisition process and how concepts are advanced and updated through a process of inference, dispute, examination and vice versa (Lin and Chan 2018). A continuing, evolutionary epistemology can be generalised to represent knowledge emergence which, in summary, is a Darwinian theory of knowledge growth; we obtain knowledge which solves our problems, and gets something that approaches being adequate for our tentative solutions (Vanberg 2014).

Nevertheless, the mere accumulation of beliefs, unless this is connected to truth, is often not desirable. Our epistemic goals aim to acquire true beliefs which combine with our realistic goals. The theory of justification acknowledges the possibility of false assumptions. Truth cannot be directly verified but provides a reason for why argument is essential. Assuming that all beliefs are necessarily true severely hin-

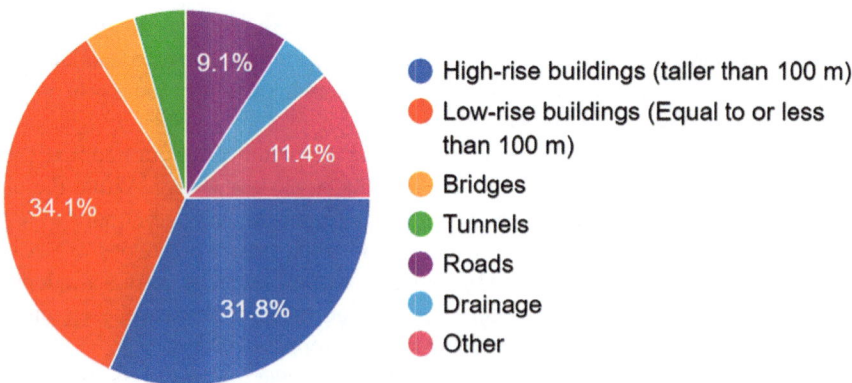

Fig. 1 Types of work whereby workers become involved in accidents

ders our understanding of knowledge in a social world. Differences and ambiguities between knowledge and information have been raised in economic literature from time to time. If some knowledge cannot be conveyed easily, this explains why it is not commonly shared (Leppälä 2015).

New ideas are generated by reconfiguring and recombining old ideas. People with diverse backgrounds interact with each other, incorporating others' opinions into theirs and creating something new as a result. Knowledge variety implies knowledge dispersed. Dispersed knowledge is seen as a challenge to its efficient use and an opportunity. Suitable institutions allow the active dispersal of expertise for use (Leppälä 2015).

The nature of why rational actors have true rather than false beliefs, why some information is likely to be wrong instead of right, and how rational actors react to uncertainty as such can be addressed by justification theory. Truth does affect not only knowledge dissemination, but also the possibility of the generation and use of knowledge and false information. Uncertainty in knowledge generation implies the possibility of an effort without results (Leppälä 2015).

4   Research Method

In this research, we compare the subjective knowledge of workers who have come across accidents and the objective views of the legal reports about what factors lead to construction accidents. Among our 46 construction injury cases in Hong Kong, about 2/3 of them involve accidents in building constructions in Hong Kong. Road construction accounts for 9.1% of construction accidents and ranks third among all the construction project types, regarding the occurrence of accidents (Fig. 1).

We also study the likelihood of construction injuries in relation to months, week-days and days. Accidents are more likely on the first day of the month, and on Mondays and Fridays. Also, the frequency of construction injuries is different according to the month. It is clear that such injuries are more likely to happen in summer and winter than in spring and autumn. In addition, the accident rate in summer is higher than that in winter. Moreover, all of the results make sense regarding quite straightforwardly likely causes. On Monday and Friday, construction workers may be thinking about the weekend, one way or another, and this leads to construction workers becoming careless. In summer and winter, the extreme weather makes construction even more difficult than at other times. So, the likelihood of construction injuries increases as well. From the graph included, we can see that, regarding parts of the body, the parts which are most subjected to injury are the hands, legs and waist. We can explain this result in two ways. First, construction workers usually use their hands and legs to finish their work. Second, it is difficult for them to protect their hands and legs by traditional means, such as the use of shoes and gloves. In practice, it is easier for them to protect their eyes and ears through protective glass and earplugs. But, there is no light, convenient and effective means of protecting the hands and legs. If they wear protective shoes and gloves, this makes it difficult for them to finish their work and even leads to further construction accidents (Figs. 2, 3 and 4).

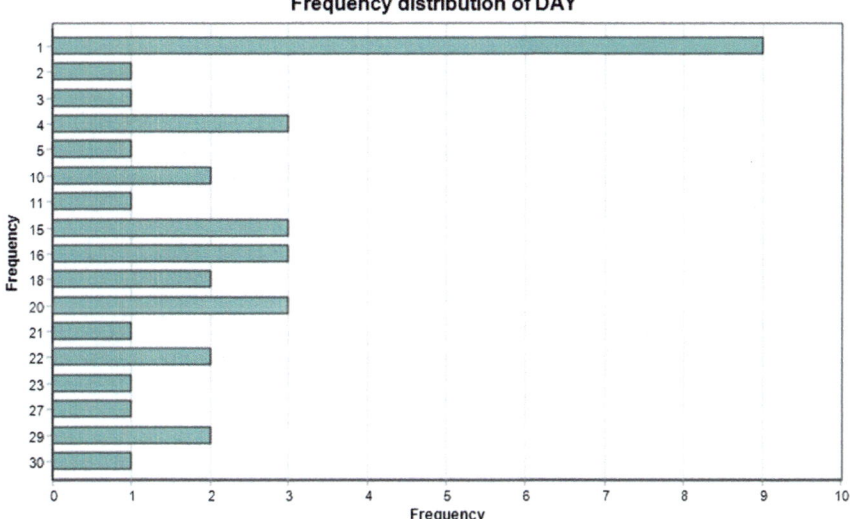

Fig. 2 Accidents mapped against the days on which these happen among construction workers

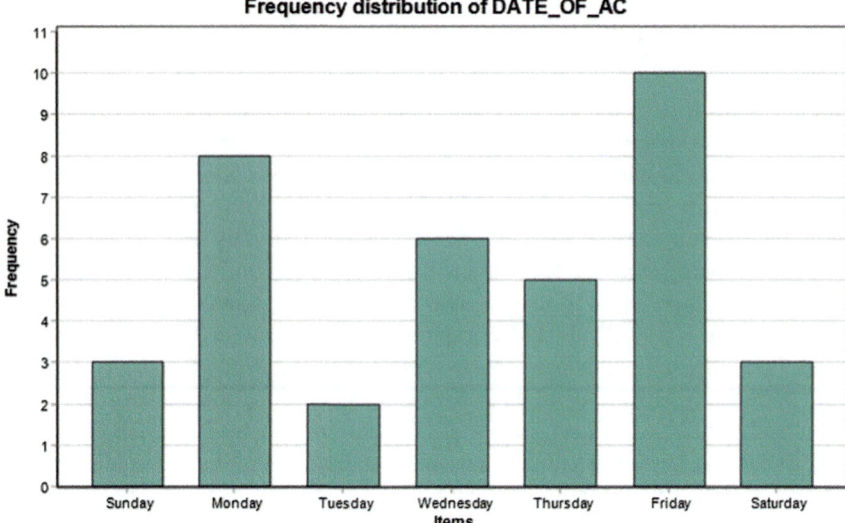

Fig. 3 Weekdays when the accidents occur

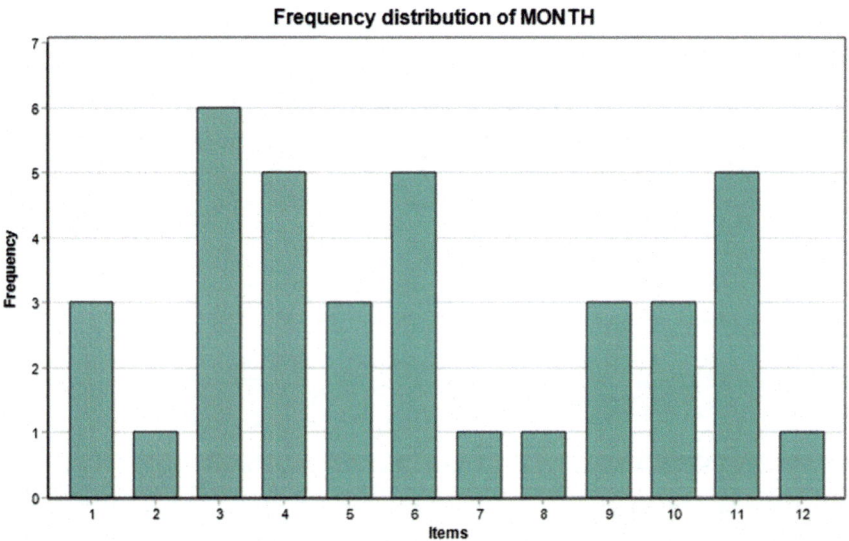

Fig. 4 Months when accidents occurred

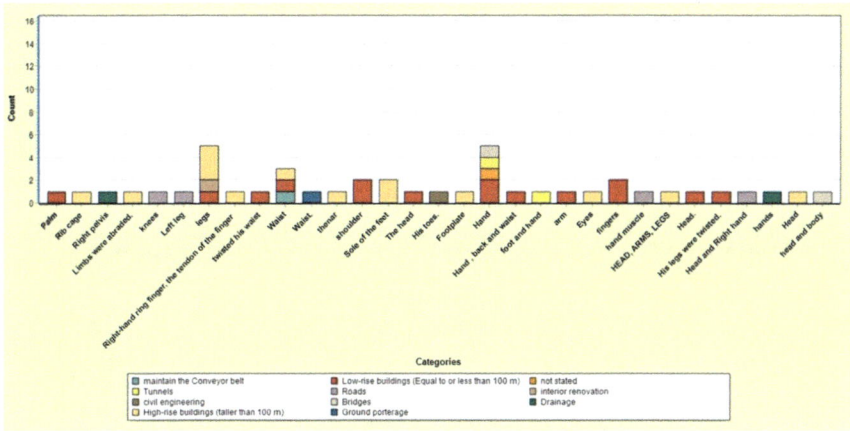

Fig. 5 Frequency distribution concerning part of body injured

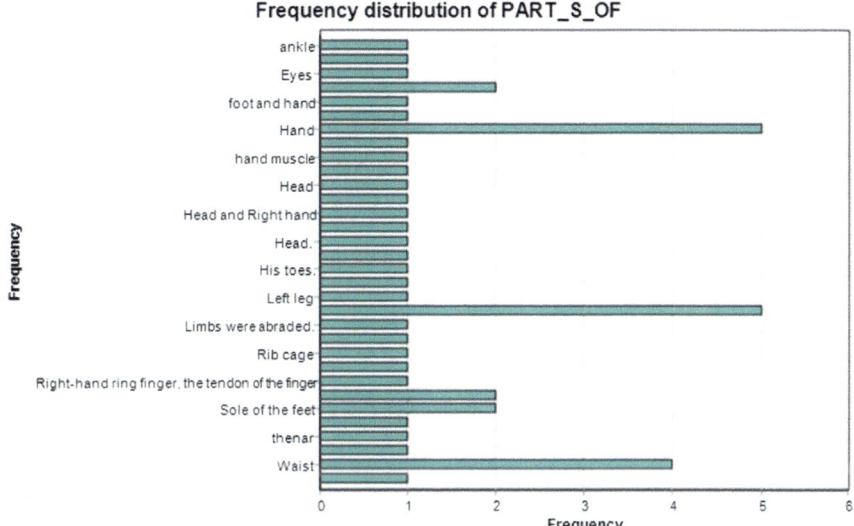

The crosstab graph shows what building projects are the most likely leading to injury, because these two construction projects are too complex to include a lot of material agents (e.g. a piece of wood which hits a workers' head is a material agent). Thus, many of the injuries which happen on high-rise sites are to limbs (Figs. 5 and 6).

When workers are asked about the factors that lead to construction accidents, the majority of them only consider their faults about the accidents, and the material agents which were involved. For example, worker 1 believes that the most important reason for an accident was that he did not wear gloves and that, therefore, his hands were cut by a wire. Worker 2 considered his incorrect position when he transported materials was the primary cause. Worker 20 suggested that since he was a junior worker at that

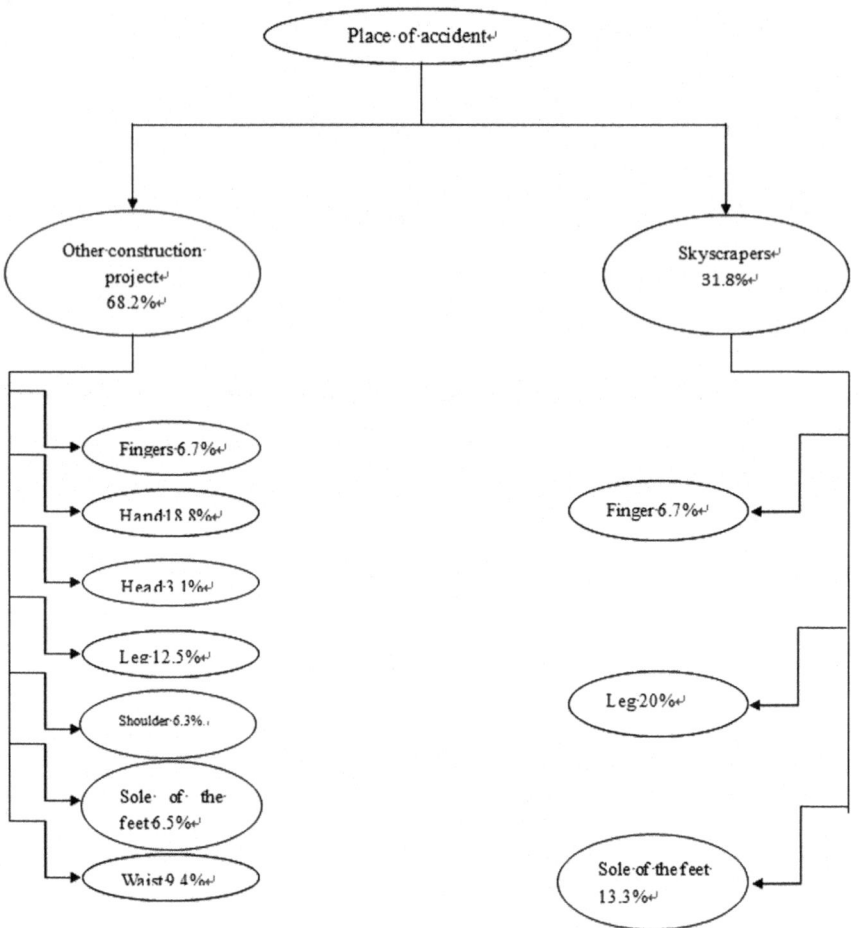

Fig. 6 Frequency distribution of part of body injured in the cases of high-rise projects and other types of building projects

time, he faced the problem of a lack of safety knowledge and experience. He ignored the importance of using the correct posture when carrying heavy materials.

One of the workers considered housekeeping to be a significant factor; interviewee 4 suggested that the site was messy with a great deal of rubbish and dangerous stuff distributed around it, and that when workers accidentally step on these different kinds of material, they are at risk of getting hurt. Without wearing safety equipment, it would be easy to fall from high places. Interviewee 14 considered rough and uneven ground as the primary cause.

Only one worker argued that the 'boss' should be the one who should be held responsible for the accidents. This is in sharp contrast to the objective court decisions as shown in the following section.

Table 1 Responsibility of different parties in construction accidents

The relative importance of contractors and employers in construction accidents	Number
Contractors 100%	28
Contractor 70%; workers 30%	1
Contractors 75%	1
Employer 50%, subcontractor 50%	1
No information	2
Workers 60% contributory negligence; 40% contractors	1
Workers 25%	1
Workers 100%	5
Workers 25%	1
Grand total	41

5 Causes that Led to Accidents: A Critical Review on Court Cases

In the second part of this research, we review the construction accident court cases from 2013 to 2015. Many of these accidents happened consisted of moving objects striking individuals or falls from height. While the previous research suggests that heat stress may be one of the causes that lead to construction accidents, this is not reflected in the cases which ultimately end up in court. October is the month in which the most significant numbers of accidents which lead to lawsuits occur. Judges mainly look at the issue from a management perspective. For example, in *Lo Wai Shing* v *Lik Sang Engineering Co Ltd* [2013] HKEC 1755, it was suggested that the defendant had failed to ensure that the plaintiff would be reasonably safe while working on a lorry because they (the defendant) had not provided a safe system of work and assured that the system was adequately supervised. In *Singh Harjit* v *Determination Business Ltd* [2015] HKEC 332, there is a strong prima facie case of failure to maintain a safe place or system of work on the part of the defendants. In *Lam Kwong Sum* v *Wong Hau Ling* [2013] HKEC 569, the contractors failed to have a safe system of work (i.e. it was common for the dumper to be overloaded, generally, and indeed it was at the time of the accident) or a safe place of work (i.e. failing to place effective barrier at the edge from where the dumper fell over the soil berm) (Table 1; Figs. 7, 8 and 9).

6 Discussion and Conclusion

Our research has found that the subjective perspectives of construction workers deviate from the judges' viewpoints concerning the causes of accidents. While most of the workers considered that accident occurred because of their faults, the judges believed

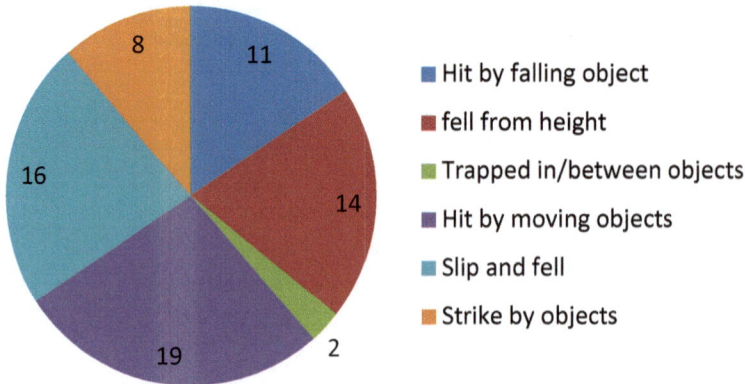

Fig. 7   Causes of construction accidents in court cases from 2013 to 2015

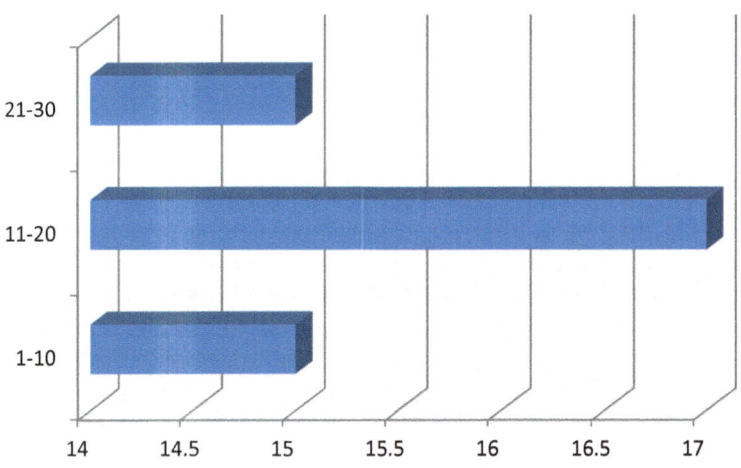

Fig. 8   Days when construction accidents happened within a month

that the employers were responsible for most of the accidents which resulted in court cases. While the literature suggests that heat stress is one of the most critical factors which leads to construction accidents, this is not reflected at all in interviews or court cases. For example, accidents which happen in October often lead to construction accident compensation cases. Regarding days of the week on which construction accidents occur, according to the court cases in 2013–2015, the chances are equal. Regarding knowledge about accidents, our paper shows that the division of labour leads to a division of knowledge. Those who come across accidents usually focus on immediate causes. Judges who have to call upon all the witnesses, experts, written documents, etc. are often concerned more about the less direct factors. As such, contractors who are responsible for managing site safety are also liable for any accidents which occur on site. This does not mean that judges will necessarily disagree

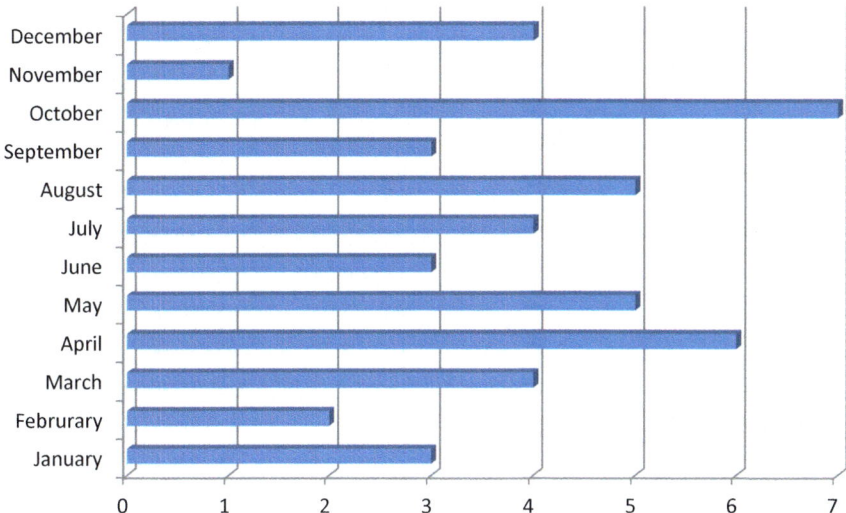

Fig. 9 Months when construction accidents happened

with the workers; they have to bear their responsibility. The proportion of negligence allocated reflects that workers are often also partly responsible for their accidents, but this responsibility is usually given a lower weight in court.

We have also seen that the mere accumulation of beliefs is often not desirable. Workers believe that construction accidents are caused by immediate causes only. These immediate causes can be removed if contractors have probably done their jobs. The knowledge of what factors lead to accidents may be to a certain extent similar to tacit knowledge, in that it is a kind of knowledge which can be challenging to share. Division of labour also divides the knowledge related to construction safety. High transaction costs in knowledge sharing hinder the actual knowledge that we can receive.

References

Amiri, M., A. Ardeshir, M. Hossein, and F. Zarandi. 2014. Risk-based analysis of construction accidents in Iran during 2007–2011—Meta analyze study (National Report). *Iranian Journal of Public Health* 43 (4): 507–522.

Arquillos, A.L., J.C.R. Romero, and A. Gibb. 2012. Analysis of construction accidents in Spain, 2003–2008. *Journal of Safety Research* 43 (5/6): 381–388.

Brogmus, G.E. 2007. Day of the week lost time occupational injury trends in the US by gender and industry and their implications for work scheduling. *Ergonomics* 50 (3): 446–474.

Gambatese, J.A., M. Behm, and S. Rajendran. 2008. Design's role in construction accident causality and prevention: Perspectives from an expert panel. *Safety Science* 46 (4): 675–691.

Haslam, R.A., S.A. Hide, G.F. Gibb, D.E. Gyi, S. Atkinson, and A.R. Duff. 2005. Contributing factors in construction accidents. *Applied Ergonomics* 36: 401–415.

Hayek, F.A. 1937. Economics and knowledge. *Economica* 4: 33–54.

Hayek, F.A. 1945. The use of knowledge in society. *American Economic Review* 35 (4): 519–530.

Leppälä, S. 2015. Economic analysis of knowledge: The history of thought and the central themes. *Journal of Economic Surveys* 29 (2): 263–286.

Li, R.Y.M. 2018. *An economic analysis on automated construction safety: Internet of things, artificial intelligence and 3D printing*. Singapore: Springer.

Li, R.Y.M., and S.W. Poon. 2011. Using web 2.0 to share knowledge of construction safety: The fable of economic animals. *Economic Affairs* 31 (1): 73–79.

Li, R.Y.M., and S.W. Poon. 2013. *Construction safety*. Berlin: Springer.

Lin, F., and C.K.K. Chan. 2018. Promoting elementary students' epistemology of science through computer-supported knowledge-building discourse and epistemic reflection. *International Journal of Science Education* 40 (6): 668–687.

Rowlinson, S., and Y.A. Jia. 2015. Construction accident causality: An institutional analysis of heat illness incidents on site. *Safety Science* 78: 179–189.

Salminen, S., J. Saari, and K. Saarela. 1990. Organizational factors and risk-taking in occupational accidents. *Journal of Occupational Accidents* 12 (1–3): 134.

Vanberg, V.J. 2014. Darwinian paradigm, cultural evolution and human purposes: On F.A. Hayek's evolutionary view of the market. *Journal of Evolutionary Economics* 24 (1): 35–57.

Yilmaz, F. 2015. Monitoring and analysis of construction site accidents by using accidents analysis management system in Turkey. *Journal of Sustainable Development* 8 (2).

Printed by Printforce, the Netherlands